Books by Kathryn Casey

DIE, MY LOVE
SHE WANTED IT ALL
A WARRANT TO KILL
THE RAPIST'S WIFE

DIE, MY LOVE

A TRUE STORY OF REVENGE, MURDER, AND TWO TEXAS SISTERS

KATHRYN CASEY

HARPER

An Imprint of HarperCollinsPublishers

DIE, MY LOVE is a journalistic account of the actual murder investigation of Piper Rountree for the 2005 killing of Fred Jablin in Richmond, Virginia. The events recounted in this book are true. The personalities, events, actions, and conversations in this book have been constructed using court documents, including trial transcripts, extensive interviews, letters, personal papers, research, and press accounts. Quoted testimony has been taken verbatim from court transcripts and other sworn statements. Some names have been changed to protect the privacy of those individuals.

HARPER

An Imprint of HarperCollins*Publishers*
10 East 53rd Street
New York, New York 10022-5299

Copyright © 2007 by Kathryn Casey
ISBN: 978-0-06-084620-6
ISBN-10: 0-06-084620-8

First Harper paperback printing: May 2007

HarperCollins® and Harper® are trademarks of HarperCollins Publishers.

Printed in the United States of America

Visit Harper paperbacks on the World Wide Web at
www.harpercollins.com

10 9 8 7 6 5 4 3 2 1

For my parents, Nick and LaVerne,
with love and gratitude.

Well, maybe there's a God above
But all I've ever learned from love
Was how to shoot somebody who outdrew you.
It's not a cry you hear at night
It's not somebody who's seen the light
It's a cold and it's a broken Hallelujah

Some names and identifying characteristics have been changed throughout this book. They include: Dr. Jim and Elaine Gable, Carol Freed, and Linda Purcell.

1

· ·

Violence can erupt at the quietest moments, in the most secure places, to the unlikeliest victims. So it was early on the morning of October 30, 2004, in the tranquil Richmond, Virginia, suburb of Kingsley. From that moment forward, lives would be changed and perceptions of the world forever altered. Family, friends, neighbors, and colleagues would understand with a new certainty how selfish and dangerous love can be.

Throughout Kingsley, massive oaks and maples surrounded impressive brick homes set back from streets that rolled with a gentle undulation. The entrance to the subdivision was marked with a prim green and white sign, and the day before Halloween, well-tended yards were replete with leaf-bag pumpkins and scarecrows fashioned of worn jeans and faded plaid shirts stuffed with brittle straw. Bedsheet ghosts wafted gently in the breeze as they dangled from the near barren limbs of trees still clutching the last remains of fall's red and gold. The following morning daylight savings time would begin, and the streets would be bright at just past six-thirty. But this morning Hearthglow Lane remained shrouded in night.

As the gunshots echoed through the quiet neighborhood, dogs barked in yards and frightened neighbors rousted from bed ran to windows, where they stared out into the quiet streets. Only a gray-haired salesman named Bob

McArdle caught a fleeting glimpse of a mysterious figure sprinting through the neighborhood, directly in front of his house. A jogger? he wondered. Perhaps. Or could this person—he couldn't tell if it was a man or woman—be responsible for the gunfire?

"Maybe it was only a car backfiring?" the 911 dispatcher asked, questioning him.

"No," McArdle, a former Marine, insisted. "It was gunfire."

Within minutes of the 911 call a Henrico County squad car snaked slowly down Hearthglow Lane, shining high-beam flashlights that threw shimmering funnels of light across lawns, onto front doors, and into the curtained windows of homes where some families lingered in bed as others gathered for breakfast, preparing for the errands and plans that awaited them that Saturday. In the darkness, three uniformed officers searched but found nothing out of the ordinary on Hearthglow, convincing them the gunshots must have originated elsewhere. They fanned out, combing the rest of the neighborhood. It had happened before, reports of gunshots that were never explained, the source never found. They must have wondered: Was there anything to look for in the early morning darkness? It seemed unlikely. Bloodshed was an uncommon visitor to Richmond's affluent bedroom communities called the West End.

At first glance the stately brick home at 1515 Hearthglow Lane appeared unremarkable. Inside the house, steaming coffee drained into a glass pot in the kitchen, while upstairs in their bedrooms, three children slept peacefully, unaware of the nightmare that awaited them. Once they awoke, nothing in their young lives would ever be the same. How could they ever forget the horror of this chill fall morning?

Outside on the long, narrow asphalt driveway, their father, Fred Jablin, lay dying, his life leaking out in a steady stream of dark crimson blood.

On his stomach with his head turned to the side, Fred's eyes were open and staring out toward the street. When he'd fallen, he landed on a thin bed of brown leaves, his head hitting a row of brick that lined the driveway, directly under his children's basketball hoop. Later observers would describe him as lying in a near fetal position, knees slightly bent, as if in death he'd tried to retreat into the tightness of his very beginnings, his mother's womb.

All around Fred the breeze whispered, a susurrus born of ruffling, withering leaves.

Of all the ways death can come, who would have predicted that Fred Jablin's life would end this way, gunned down in his own driveway? He was the most improbable of victims.

The world knew Fred as an esteemed University of Richmond professor, a man who lectured to thousands across the world, whose ideas helped define the field of organizational communication. His work and his life were based on logic and painstaking attention to detail, and his personal habits were regimented. A meticulous man, he kept to a precise routine. It was that predictable schedule that had made him so vulnerable: Those close to him knew that each morning at approximately six, he awoke, put on coffee to brew, then walked not out the front door, but the back one, near the kitchen, to claim his newspaper off the driveway.

Perhaps this morning he had smiled as he emerged from the house, anticipating the pleasure of the Saturday that lay ahead: the neighborhood's annual pumpkin festival, an afternoon with his children, a time to play and feel young again. Did he see the figure emerging from the shadows as he shuffled outside in slippers, navy blue sweatpants and sweatshirt? As he looked down the barrel of the gun, was he surprised? Or had he often feared this might happen?

Did he plead for his life or turn to run, desperate to escape? Did the intruder say anything as the trigger was pulled? Did Fred Jablin cry out in horror into the pitch-dark night?

It happened so quickly, life changing on a dime, as they say. One moment Fred was planning his day, anticipating all that lay ahead—then three gunshots, and suddenly he had no future. What went through his mind in those final, brief moments as he lay dying? Did postcards of his fifty-two years flash before him? Did he wonder how his world had gone so terribly wrong?

Or, as his consciousness faded, did Fred Jablin pray, entreating God to keep his children from finding his lifeless body? In his last moments of life, perhaps he simply replayed over and over again a question he'd asked so often, a question to which many doubted he'd ever found an answer: Why?

2

∗ ∗

The halls of CMA, one of the three buildings that comprise the Jesse H. Jones Communication Center at the University of Texas, Austin, were deserted and silent one evening in 1981. Students had departed hours earlier, leaving only scattered faculty quietly poring over papers in their small cluttered offices. Just after nine the silence was broken by a gruff yapping that carried through the corridors.

"Arf arf arf arf arf," Fredric Mark Jablin, an assistant professor, barked as he began hurriedly stowing the day's work in his briefcase for the night. Some said Jablin, a wiry man with a quiet manner, spare physique, glasses, and receding hairline, resembled a young Woody Allen. A serious nearly thirty-year-old who'd already made his mark in the academic world, Jablin had a playful nature, a sharp sense of humor, and a zest for living. He'd been the one to start this nightly ritual, a canine call to arms that alerted colleagues it was time for a bit of merriment.

From down the hall came an answer: *"Woof woof,"* assistant professor John Daly barked back, brusque and loud. Tall, with a warm manner and intelligent, intense eyes, Daly looked up from his work and smiled, then glanced at his watch. Laughing softly, he closed the books he'd been combing through, stacking them in a pile for morning.

Just then Jablin peered in Daly's office door. "Ready to go?" he asked with a wide grin.

"Let's be off," Daly said with a flourish.

Moments later the clacking of their heels filled the hallways as they hurried out of the building, then across the street, eager for a beer at the nearby Hole in the Wall bar.

As hard as Jablin and Daly worked throughout that long day and into the evening, teaching and compiling data for their research, they played equally hard at night. It was an exuberant time in both their lives, one filled with promise and expectation. They'd studied long and hard, and now they were at the beginning of their careers, John in interpersonal communications, while Fred was a rising star in a burgeoning field: organizational communication, the study of how people within corporations and organizations interact. While it may have seemed dry to many, to Jablin it held excitement and great promise, giving him the opportunity to explore what made some associations work while others failed. The implications were invaluable in business, education, even, at its very core, in life.

The best of friends, Jablin and Daly had first met during graduate school, while working on doctorates at Purdue University in West Lafayette, Indiana. While fields like drama and literature are centuries old, organizational communication was little more than a toddler at the time, still in its first quarter century. Jablin was one of the field's rising stars. Within a year of earning his Ph.D., he'd caused a stir that reverberated through his field when he compiled a benchmark paper, one that reevaluated all the existing literature.

"Fred had a real talent," says Professor Lyle Sussman, Jablin's master's degree supervisor at the University of Michigan. "He could take mountains of research and sift through it until he saw overriding themes. Then he drew conclusions, unique ideas and concepts."

More than that, his research was meticulous, well-grounded, and followed established scientific methods. This preciseness came naturally; Fredric Jablin grew up wanting to be a scientist. As a boy in the Jewish working-class Floral Park, New York, suburb—the son of Irving Jablin, an accountant, and Mildred, a bookkeeper—he watched a neighbor who worked as a scientist leave for work each day in his white lab coat. "I wanted to do that," Jablin later told friends. "That's how I saw my future."

The Jablins were a close family. Irving was a quiet man, and Mildred, a petite yet robust woman, served gefilte fish with purple horseradish, and on holidays her special chopped chicken liver paté. Fredric was born in 1952, the year Dwight Eisenhower was voted into the White House. It was a simpler time. Television was a new addition to America's living rooms. *Romper Room* debuted when he was just two, urging children to "clean their plates," and on *Howdy Doody*, Oil Well Willie sang songs while Clarabell honked his horn at Buffalo Bob.

In this warm, loving household, Fred was the second and youngest child, a few years behind his big brother, Michael. While they were young, Mildred stayed home to care for her boys. Mildred and Irving had lived through the Depression, and the Jablins weren't extravagant people. Yet once a month Mildred splurged, taking their sons by train into the city, to a Broadway show or to the symphony. Later in life, Fred would talk fondly of those memories.

In high school he wore that white coat he'd so coveted, when he worked in a reverse engineering lab, helping to tear products down to see how they worked. Mildred and Irving were proud of both their sons. "His parents were hardworking people," says a friend. "His mom was the quintessential Jewish mom, who would have done anything for her kids. Like Mildred, Irving was dedicated to the children."

After high school, Fred left Floral Park for Buffalo and SUNY, the State University of New York, where he majored in political science and speech. To help pay his costs, he worked as a janitor and in the school cafeteria. A brilliant student, after earning his bachelor's degree, Fred entered the University of Michigan in Ann Arbor, where he worked on a master's degree in communication. Along the way he met Marie, a woman of Italian descent from New York City. They married in 1974, when Fred was twenty-two years old, and she went with him to Purdue University, where she took graduate courses in audiology and speech pathology. He became a protégé, working on a doctorate under the professor considered the father of organizational communication, W. Charles Redding.

At Purdue, Fred studied hard and coached other graduate students. In almost every situation, he displayed a finely tuned sense of right and wrong, of morals and ethical codes. At times, friends would say, he had such a strong sense of living by the rules that he could be rigid, even to his own detriment. When a student he believed in didn't pass his comprehensives, Fred not only tutored him, but argued fiercely with the student's supervising professor, urging him to help the student pass.

Within his own career, too, Fred set up parameters he expected his mentor, Redding, to adhere to. Redding was well-known for delaying the graduations of his doctoral students, forcing them to work as much as a year longer on his research, a not unusual practice. Fred's fellow students were amazed when he had the audacity to ask Redding to sign a contract detailing precisely what Fred would have to do to graduate on time. The eminent scholar, no doubt impressed by the young man's chutzpah, agreed. "It was vintage Fred," says Daly. "He thought he knew how things were supposed to be, and he wasn't afraid to stand up for himself or others, to fight."

Later, many would describe Fred's focus on how things should be done as nearly obsessive. "Fred didn't like irrationality," says Mark Knapp, one of his professors and colleagues, who became a friend. "He believed that justice and reason should prevail in the world. The problem is that life isn't always that way."

At Purdue, Fred's office was two doors from Daly's. Often, Daly looked in and found his friend working late into the night under a green lawyer's lamp. Fred wasn't one for socializing, instead spending every available moment away from the university with Marie. Daly would always remember how Fred channeled all of his attention on work and his marriage. "When Fred was in a relationship, he was devoted," he says. While other graduate students partied on Friday nights, Fred and Marie stayed home in their campus apartment and watched *The Rockford Files.*

Still, the marriage ended, just three years after it began. Later, Fred would call it a "no-fault" divorce, an amicable and mutual parting. He'd say that they both knew the marriage wasn't working, that they wanted different things in life, and that they made the decision to end it before they had children and property to muddy up the split. Yet, it was a painful defeat for Fred. "He wasn't devastated, but he was hurt by the divorce," says Daly. "He was saddened that it hadn't worked out."

That same year, 1977, Fred earned his Ph.D. in organizational communication and moved to the Midwest to take on his first teaching position, as an assistant professor at the University of Wisconsin at Milwaukee. It was there that he wrote his breakthrough work, an article on supervisor and subordinate communication that was accepted for publication in the prestigious *Psychological Bulletin.*

"At that point, organizational communication was just getting its legs," says Mark Knapp. "Getting an article published in another discipline's leading journal was a coup. With that paper, Fred leapfrogged to the head of his class and built a reputation for himself in the field."

Shortly after, Fred received an offer from the University

of Texas at Austin, where Daly had landed a teaching position. Enthused about the big name, high-profile school and about working with his old friend, Fred packed everything he owned, including his yellow Toyota Corolla, in a rented truck, and drove from Milwaukee to Texas.

With its rolling hills, Austin is a welcoming place, large enough to offer the pleasures of a city but small enough not to impinge on the enjoyment of life. Downtown, the skyline was dominated by the University of Texas clock tower—lit a burnt orange, the school color, when the university's teams win—and the dome of the Texas State Capitol. Austin has the feel of a place where dreams can come true, and the big Texas sky with its vast horizons seems to promise all that is possible.

By the time he'd reached UT, Fred Jablin had realized his youthful dream; he didn't wear a lab coat, but he was a scientist. Instead of chemicals in test tubes, he dissected the ways people interacted and how they affected success in the workplace. At UT he videotaped job interviews, then, armed with his observations, scrutinized the communication between recruiters and would-be employees. He researched the dynamics of brainstorming and the socialization of new employees into companies. In the end, Fred Jablin diagnosed how and why things happened between people in business, explaining how corporations could interview more skillfully and increase the odds of hiring the best candidate. Always, he worked hard. During the first five years of his UT career, he published an impressive thirteen articles in scholarly journals. Through it all, Fred's enthusiasm was infectious. Decades later, many would remember how excited he'd get at the prospect of a new idea, a new theory.

Of the two in the Austin bar that night, where they'd gone after work, Daly was the tall, handsome one, with a commanding voice and presence. Physically, Fred was more average looking than handsome. At one time he wore a straggly toupee to cover his balding pate, saying he did so to please his mother. When he drove himself to the hospital for an emergency appendectomy, it was lost in the fray, and to the delight of his friends, he didn't replace it. Still, they saw him as somewhat eccentric. As quiet and intense as he often seemed—almost painfully so at times—Fred talked with a rapid-fire New York accent.

"Fred was always a bit of a nebbish," says an old friend with a laugh. "But he had a great smile, an ear-to-ear grin, and an infectious curiosity that made him stand out in a crowd."

"Once you knew Fred, with his wry sense of humor and his contagious love of life, nothing else mattered," says Daly.

In Austin, Fred paid $50,000 for a 1,500-square-foot house on Harper's Ferry, a quiet residential street, that he decorated with auction finds. He loved the excitement of bidding and enjoyed getting a good deal. Quickly, he carved out a social life. Halloween was big in Austin, and Fred covered the house with cobwebs, spiders, and carved pumpkins, and posed Mr. Meyers—a dummy with long, bony fingers, dressed in a flannel shirt and blue jeans—as if it were a guest at his annual party. "Fred believed that Halloween was a time when you could be a different person," says Daly. "He'd spend days making sure the decorations were just right, arranging and rearranging the cobwebs so they looked real."

Along with adorning the house, Fred obsessed over his costumes. The year of the Extra Strength Tylenol poisonings in Chicago, he wore a giant pill bottle with the painkiller's label affixed to the front. Another year, he was a black and yellow bumblebee. His favorite was an elaborate wizard costume. On the nights he portrayed a Merlinesque sorcerer, he walked around the party, waving a wand and casting make-believe spells on his laughing guests.

"Austin's a liberal college town, and a few people at the

party smoked pot. Most of us stood around and talked shop," says Knapp. "Fred's parties weren't wild, but they were fun. Everyone, including Fred, had a good time."

In fact, for all his eccentricities and his dedication to hard work, Fred had a real knack for enjoying life.

In the summers, he and Daly rented Sunfish catamarans to sail on nearby Lake Travis, a spectacular setting surrounded by jagged hills and a tall pine forest. One year, they took flying lessons and went soaring in gliders. Years later, Daly would recount an incident one night while the two friends were scuba diving. Fifty feet down in the lake, Daly was spooked when an enormous catfish, "one that looked bigger than a whale," swam into his flashlight beam, then directly at his face, staring at him through his mask. Daly panicked and leapt up to flee to the surface. Fred pulled him back down, turned off the flashlight so the fish would leave, and calmed his friend, so he wouldn't surface too quickly and risk having bends. "He may have saved my life," says Daly.

The things Fred Jablin loved, like his gas-efficient Toyota and reading the *Wall Street Journal*, weren't elaborate. More than anything, he enjoyed his friends, his work, and the small pleasures of life. Fred and Daly's barking salutations before they left their offices became a tradition, yet it was a secret they shared with only a choice few, and they barely restrained themselves the day an elderly professor confided in them, "You know, at night, someone brings their dogs up here." When the old man left the room, John and Fred erupted in boyish laughter.

Evenings at the Austin bar were filled with merriment. They talked and drank beer, comparing their days and feeding off each other's ideas. They looked as if they hadn't a care in the world, and perhaps they didn't. It could have been argued that there was little that could have made the two friends happier than what they were doing. They had their differences: Daly was married, and he'd expanded his career, taking on lucrative consulting jobs with corporations. Fred, single since the divorce from Marie, was resolutely devoted to his research and teaching. "Fred was a true teacher at heart," says Daly. "There was nothing he enjoyed more than working one-on-one with a bright student."

John Daly introduced Fred to one such student in the fall of 1981. Actually, Piper Ann Rountree had been in Fred's communications class the previous year, but it was a large lecture hall filled with more than a hundred students. Later Fred would say he didn't remember her, except that she was the girl who once brought her mother to class and introduced the older woman to him after the session.

By that fall, however, Piper, a senior with an inquisitive mind, was frequenting the faculty offices in the communications building, hanging around, as some students were wont to do, talking to her professors between and after classes. That was how Daly and Piper struck up a conversation. From that moment on, she was a frequent presence in the beige-walled and cluttered faculty offices and conference room. Later, many would remember her, for Piper Rountree was the kind of young woman who made an impression. She had a quick mind and she was beautiful, with a petite, lithe, athletic body, shoulder-length dark brown hair, and intelligent dark eyes.

Years later Daly would recall how he mentioned to Piper one afternoon that he and Professor Jablin had an event planned for the coming day. She asked to go along. He agreed. When the day arrived, Daly was busy and needed to cancel. With no way to reach Piper, he called Fred. "We're supposed to meet this student, Piper Rountree, in the parking lot," he told him. "I can't make it, but why don't you go ahead and go with her?"

"Sure," Fred said. "No problem."

Something happened in that parking lot that day when

Fred Jablin met Piper Rountree. He was eight years older than she, and they'd grown up in very different worlds. Yet there was an immediate attraction.

"That was the beginning. From that point on, Fred and Piper were infatuated with each other," says Daly. Reflecting on how that single event changed his friend's life, Daly paused before adding, "That will always be with me, that I was the one who introduced them and started it all."

3

They were such a strange match that many would question why Fred Jablin and Piper Rountree bonded so quickly and so tightly. Fred was precise and exacting, at times even unyielding, but a calm and measured presence. He believed the world was fashioned out of black and white, with little room for shades of gray. Piper, on the other hand, was a free spirit, ethereal and gossamer, like a butterfly flitting from flower to flower. Volatile and with a quick temper, she was the quirky hippie, the young, exuberant, artistic earth mother who rescued stray animals. Fred was a plain man, whose clothes hung loose and rumpled on his slight frame, with little more interruption than they might on a hanger. With her classic features, Piper was truly beautiful in her tight jeans, boots, and peasant blouses. Together, they were the quintessential odd couple, the epitome of the old adage: opposites attract.

"We both loved to have fun, and we had fun together," Piper would say many years later, trying to explain the instant attraction. "In his own way, at least back then, Fred could be playful, like at Halloween. We both liked to dress up, to pretend to be something we weren't for the night. We had that in common and other things. At first, well, at first it was good."

At first there were summer days spent at Lake Travis, including forays to Hippie Hollow, Austin's clothing optional beach, with its rocky shoreline. Stripped of all inhibitions, Fred playfully pounded his chest like Tarzan and jumped into the chilly water. They double-dated with Daly and his wife, Chris, evenings spent over dinner or at the movies. Soon, neighbors on Fred's quiet street began seeing a comely young woman jogging in the mornings.

"Fred introduced us, and we liked Piper," says Leo Kuentz, who, with his wife, Linda, and their children, lived next door to Fred. The Kuentzes and Fred had moved into the neighborhood just months apart and became fast friends. Leo worked as a repairman for the local telephone company. He and Fred were so close, they didn't put a fence between their yards, and circulated back and forth.

Although very different, the two men had a lot in common. They were both into finely tuned stereo equipment with big speakers—although Leo, a large, strongly built man, listened to rock and roll and Fred preferred jazz. They worked on projects around the houses together, and they talked. Leo enjoyed listening to Fred chatter about his work and his theories on life. Somehow, Fred always seemed able to boil everything down and explain it, often in terms of a prevailing theory of communication. "I'd take a customer relations class at work and tell Fred what they were teaching us," Leo recalled years later, "and Fred would say, 'You know, Leo, that's based on so-and-so professor's work. That's the theory of this or that.' Fred loved what he did. He lived it."

Linda and Piper didn't hit it off quite as well. It wasn't that they didn't get along, but that they didn't develop the closeness the Kuentzes had with Fred. "Piper was fun and pretty and full of energy," says Linda, a petite woman with blond hair. "It was just that we never got really tight. What I remember about Piper is that she seemed closest to her family. She talked about the Rountrees all the time, especially Tina."

Many would remember the Rountrees that way, an exceptionally close family, warm and inviting. "They were a real presence in Harlingen," one family friend would remember. "They seemed happy to be with each other, and people in town felt privileged to know them."

Piper was the youngest of the five children of Betty Jane and Dr. William Coleman Rountree, Jr. Betty had studied to be a nurse, and Bill was an Air Force surgeon, a heart specialist, and the son of a doctor before him. He had earned his M.D. at Tulane University Medical School in New Orleans after World War II, and the children were born all over the world; the oldest, William III, in New Orleans in 1948, and Piper, the youngest, twelve years later, on January 6, 1960, in Japan. In between there were three other Rountree children: Tina, Jean, and Tom. When asked about her unusual name, Piper would claim that she'd been named after a bottle of France's Piper-Heidsieck champagne.

The Rountree family went back at least three generations in South Texas, and, according to Piper, her paternal grandfather married five times, once to a sister of the infamous outlaw Jesse James. Her paternal grandfather had constructed a family tree, and legend had it that he was a descendent of Moses Rountree, an orphan discovered in England during the 1300s under a rowan tree, a species believed to have mystical powers to ward off witches and evil spirits. Believers carved crosses and walking sticks from the wood to use as protection. Throughout the years, Piper talked of her heritage often, describing being a Rountree as "important."

"Since we moved every year or two, we were all really close," says Tina. "We came to rely on one another more than on anyone else. We were all best friends."

Perhaps that closeness developed at least partly because of the family troubles. Years later Piper would describe their father, William, as an alcoholic. "I barely knew him," she'd say.

"First he was drinking, and when he drank he could be mean. Then, when I was still little, he was out of the picture."

William Rountree retired from the military in the mid-sixties, while the family was stationed in Harlingen, Texas, and most of Piper's growing-up years would be spent there. Harlingen, a small city on the southern tip of Texas, thirty miles from the border, lies in what Texans call "the Valley," a meld of the U.S. and Mexico populated by ranch houses, palm trees, and miles of mobile homes and citrus groves.

The city, which boasted little more than 41,000 residents at the time Piper was growing up, has a rich history. It began in the mid-1800s as a frontier town called "Six-shooter Junction," after the Texas Rangers and border patrol officers stationed there. Irrigation canals off the Rio Grande River fed the crops and gave Harlingen its name, christened after Van Harlingen, in the Netherlands, another town located on a coastal plane and crossed by canals. The predominant race and culture of the area is Hispanic, and nearly a third of the city's residents fall below the poverty level. "Harlingen's always been a sleepy little city," says one of Piper's childhood friends. "Everyone's goal was to graduate from high school, go to college, and move away."

Looking back, it would seem that Piper's childhood was plagued by trauma. In September 1967, when she was seven years old, her father took her to the beach at nearby Padre Island following Hurricane Beulah, a category four storm that ravaged the South Texas coastline with waves of tornadoes. While there, she stood under a damaged house on stilts. "The house fell on me," she says. "The foundation gave way, and it just collapsed." She suffered broken ribs and a ruptured spleen that had to be removed. Afterward, she was prone to illness, and she suffered nightmares, reliving the terror.

Then, when Piper was nine, her father suffered a stroke. William Rountree was just forty-eight at the time. "He recovered, but he continued to smoke and drink," says Piper's oldest sister, Tina. Afterward, Piper would say he was meaner than before, and her mother, Betty, would say that her husband had a "personality change" that forced him to be moved to a mobile home the family owned near the beach. Many of Piper's friends would grow up believing that Piper's mother was divorced from her husband, since Piper's father was absent for much of her school years.

After her husband's stroke, Betty took up real estate. She grew a good business, and Piper's childhood friends would remember her as a solid presence. "Betty was a cool mom, the kind we all wanted," says one of them. "She was strict, but she talked to us, not down to us. She went to PTA meetings and did all the stuff. She was always there for her kids."

"Piper was the baby of the family, and my mom was really protective," says Tina. Piper's friends remember Betty fawning over Piper as a little girl, showering her with attention. That wasn't hard. Piper was a bright, attractive child. With thick dark hair, a compact little body, and a delicate face, she was artistic, a talent she'd later claim she inherited from a grandfather, who was a structural engineer. Betty would later say the one time Piper got in trouble was when she and her friend Lisa found the family wine cabinet and drank until they both became ill. With a physician father and a realtor for a mother, Piper enjoyed a certain status in Harlingen. "The Rountrees weren't rich, but they were socially up there," says one of Piper's friends. "They were clannish, kind of a tight-knit family, but they were well thought of."

Despite what she'd call a "good childhood," Piper would describe her relationship with her mother very differently. Years later she'd tell a psychiatrist that Betty was absent much of the time: "It was really Tina who raised me. Tina was always more of a mother to me than my mother."

Although eight years separated the two girls, they were

extremely close. Piper's friends would remember that when they were in high school, Tina, who'd already graduated from college, went out with them at night to a park where the teenagers smoked, drank, and talked. "They were closer than any sisters I've ever known," says one of Piper's friends. "It was obvious that Tina was there for Piper, and Piper was there for Tina. Everyone understood that."

"I admired the way Piper was so popular in high school. I never was," Tina said, her blue eyes flashing. "The thing is that Piper and I are closer than ordinary sisters. We're soul sisters. We're incomplete without each other. As children, my mother made me sleep in bed with Piper. Even as adults, we frequently slept together. We were so close, it just felt right. I love her so much, it was like we were part of each other."

In school, Piper was "on the A-list," says one classmate. "She ran with the rich crowd, and she was one of *those* girls, the ones who seemed to have everything going for them." Some would call her quirky. She dressed in striped knee socks, layers of mismatched clothes, and big chunky shoes before they were in style. She was a top student. In 1978, the year she graduated, the Harlingen High School yearbook, called *El Arroyo*, translation "The Stream," listed Piper's many accomplishments: National Honor Society, student council, speech, drama, tennis team, editor of the school's literary publication, homecoming court, choir, yearbook committee, and third place at the science fair. Many would remember Piper as a bookworm, a student who devoured the school library, reading Aristotle and Milton.

"Anything she ever did at school, she succeeded in," says a friend.

With friends, Piper was giving, a motherly young girl who helped others, volunteering at the hospital, pitching in to tutor her friends who struggled with their classes. And before it became fashionable, Piper was dedicated to the environment. From an early age, she worried about the effects of the ozone layer and the plight of animals. She was so gentle and so involved with nature "she'd pick up a spider and carry it out of the house instead of killing it," says an old friend.

Despite her accomplishments, boys never seemed intimidated by Piper. "She'd start talking to them, and she was so different, so unusual, they'd become fascinated," says a friend. "She seemed to really clue in to guys, and it seemed like, if she'd asked, they would have given her anything she wanted."

Later, after everything happened, two incidents from Piper's high school years would resonate for her classmates. The first was a drama club production she was in as a student, *The Rope Dancers,* a Tony-nominated play from 1958 written by Morton Wishengrad, about an Irish-American couple whose daughter was born with six fingers. Piper played opposite a friend, John Speer, and in the play Speer's character pleaded with Piper's, asking her to show love for their daughter. "Kiss her and show her your love," he implored. Piper's character, the shrewish wife, refused.

"In the play, Piper's character couldn't love her child," Speer would recall much later. "In real life, it would turn out that perhaps Piper loved her children too much."

After the play, a friend told her she'd put on an incredible performance. Piper's comment in response would also be recalled years later, only in starkly different terms. "Piper said she liked acting because it felt good to dress up in a costume and pretend to be another person," says the friend. "I asked why she would want to be someone else, and she said it was fun to fool people into thinking she was someone she wasn't."

The other incident involved a classmate who had a serious car accident but emerged without a scrape. "You had an angel on your shoulder looking after you," Piper told the girl.

At first blush it seemed an innocuous thing to say, something anyone might utter when a person escaped what appeared to be certain injury or death. But Piper went on, explaining that she believed angels were present in the world, controlling events, and that if they were listened for, they could serve as life guides. "We all have angels we live with every day."

She was so serious, seemed so literal about it, that the friend remarked to another, "I don't know what's wrong with Piper."

"The thing about Piper was that she believed everything she did had a purpose," says another friend. "And that she literally had an angel with her at all times, looking out for her."

Perhaps the crystals Piper carried in her pockets made her feel more connected to her angel. Or maybe her belief in angels harkened back to the time the house collapsed on her. It's not unlikely that someone told her she survived because an angel was on her shoulder, protecting her. However the belief began, it would be a connection Piper felt throughout her life, this tie to another, less physical world.

In 1978, when Piper reached the University of Texas in Austin, she was following a family tradition. Tina and their brother William had both gone to UT before her. By then all the Rountree children had spread out and started their lives. William was an attorney in Harlingen, and Tina was a registered nurse who'd married an attorney and settled in Houston.

For the first three years, Piper roomed with Lavon Guerrero, a close friend from Harlingen. Lavon would talk fondly of those years, and how nurturing her old friend could be: "In the mornings, Piper would bring me tea in bed, and we'd talk. She was gentle. She baked bread, and she loved animals."

At UT, Piper majored in speech. Already fluent in Spanish from growing up in the cross-cultural environment of the Texas/Mexican border, she studied German.

In the fall of 1981, Piper's senior year at UT, Lavon had graduated and Piper was sharing the apartment with Maureen Bemko, a fellow student with whom she'd worked on the UT yearbook. Piper paid little attention to the apartment, and when Maureen moved in, she spent a week cleaning after she found the place roach-infested and dirty, with piles of dishes in the kitchen sink. Friends asked Bemko why she lived with Piper, who they saw as odd. "No one could tell me why they felt that way," Maureen would say later. "They just described her as 'different.'"

Even her old friends from home wondered about Piper. One ran into her at UT. She and Piper had been on the Harlingen High School tennis team together, and they'd both continued to play in college. The friend didn't like to ask Piper about her angel theories or her belief in crystals, "because she'd just go on and on about them." This particular day, the friend was suffering from tennis elbow. Piper, who seemed to have a crystal for every purpose, pulled one from her bag and placed it on her friend's elbow. "We always laughed at Piper because she was so small, quirky, really cute," says the friend, "but it was strange. My elbow stopped hurting. I really was cured."

Maureen saw Piper as worldly, more so than she. That view was reinforced the day two cases of German champagne were delivered to the apartment for Piper. Then there was the conversation that made Maureen stop reading a book one night.

Devoted to tennis and running every morning, Piper had the body of an adolescent girl, and on campus she wore bill caps over her short dark hair, jeans, and T-shirts. She was thin, little more than 100 pounds, and small-breasted. "Maybe I'll get breast implants," Piper said nonchalantly that day. For the next week, she wore prostheses in her bra. At the end of the experiment, she shrugged her shoulders and announced, "I don't think it's worth it. They just got in the way."

Not long after, Piper asked Maureen if she thought it was acceptable to date older men. Maureen said she didn't know, but soon Piper brought Fred Jablin to the apartment. Afterward she asked Maureen if Fred looked familiar. "He might," Piper said. "He's a professor. One of *my* professors. But don't tell anyone we're dating."

Once Piper began dating Fred Jablin, she spent less and less time at the apartment. Maureen sometimes saw Fred dropping her roommate off and picking her up in his yellow Toyota. Piper leaned over and kissed him before she left. Daly and others on the faculty knew of the relationship. "We didn't think anything of it because Fred wasn't Piper's teacher anymore," says Daly. "It wasn't really improper."

By December 1981, Piper had moved out of the apartment she'd shared with Bemko and into Fred's home on Harper's Ferry. She lived there throughout the spring semester. The Kuentzes saw her often, jogging in the neighborhood or playing with the menagerie of animals she moved into Fred's house and yard. She had a Siamese cat, and she bought Fred a rabbit. Before long she had pet ferrets and a bird. When he talked about the animals to Daly, Fred shrugged his shoulders and smiled, appearing delighted with the happy clutter Piper Rountree had brought into his life.

It seemed, too, that he knew how to make Piper happy. As he'd once barked through the UT hallways, he now barked in the backyard. "Fred could get every dog in the neighborhood barking with him," Piper would later say. "He made me laugh."

That spring, 1982, Piper graduated, with a bachelor's degree in communication and a second major in German. Afterward she left for Germany on a one-year scholarship, studying at the University of Mannheim. Once she was gone, Fred seemed lost. As he had with Marie when they were married, he'd focused on the relationship, making it, along with his work, the centerpiece of his life. When he heard that

Piper had met a man overseas, Fred flew over to reclaim her. "It was a tumultuous relationship," says Mark Knapp. "It was Fred wanting Piper and being willing to work to make her want him."

When Fred returned to UT, he seemed placated, more sure that at the end of her classes in Germany, Piper would be returning to Texas and to him.

Looking back, John Daly would say that Fred and Piper each found something in the other that they lacked: "Piper rounded Fred out, made him more of a person. With her quirkiness and her pets, she opened up a different world to him. In addition to stability, Fred brought financial support and a firm commitment to Piper. A steadfastness that she didn't have before him. Fred loved her spontaneity, her enthusiasm, that she was beautiful and had an amazing exuberance for life."

"Fred took pleasure in the relationship," says Margaret Surratt, a UT executive assistant who worked with Fred and Daly in the communication department. "I think he sometimes looked at Piper and pinched himself. Sometimes he thought that she was too good for him, that she was so beautiful, and he was lucky to have found her."

When Piper returned to the States, the relationship was in full force. No one who saw them questioned that it was serious. "They seemed very much in love," says Surratt. "They were great friends, who enjoyed being together."

In 1983, Piper moved back into Fred's home, the ranch house that shared a yard with the Kuentzes. Piper, in many ways, became Fred's world. He loved her, and he worried about her. When Linda Kuentz remarked to Fred that Piper seemed to have it all together, Fred corrected her. "Piper's really not confident at all," he told Linda. "It would surprise you how insecure she truly is."

In the fall of 1983, Piper entered St. Mary's University Law School in San Antonio. That October, on the thirteenth,

Fred and Piper married in Austin, in the living room of his home. It was a quiet wedding, with just family and a few friends. Afterward, they threw a small reception at Austin's Hyatt. Two weeks later John and Chris Daly hosted a reception for twenty-four colleagues in their home to celebrate Fred and Piper's marriage. Fred was thirty-one and Piper was twenty-three. Margaret Surratt would later remember looking at Piper that day and thinking she seemed very young.

After the party, Piper and Fred gave the Dalys an inscribed four-leaf clover to thank them. "That was a Piper thing to do. It was something that in the past never would have occurred to Fred," says Chris. "She'd really changed his life."

4

Marriage doesn't always bring with it imagined comforts: love, companionship, security, and joy. Sometimes, even before the happiness of family and friends gathered to celebrate and witness vows has time to fade, windows open into a person's psyche, hints that a new spouse may not be the person imagined. Yet, how does one walk away without trying to salvage the relationship, if love still joins the couple together?

It would later seem that this was the situation of Fred Jablin, when soon after the marriage he discovered that his new bride may not have been all she seemed.

It wasn't until after the marriage, he'd later say, that he learned Piper had been diagnosed with attention deficit disorder. Rather than a form of Ritalin, she'd later claim that her physician prescribed Ionamin, a drug most often used for dieting under the generic name phentermine. Many people have ADD and live happy, full lives, but there was more; Piper apparently had more baggage from her childhood than her upbeat demeanor let on. She'd consulted therapists to reconcile issues from her childhood, and, Fred would say, "I found out that Piper was bulimic and had been for years."

Still, he was dedicated to her and the marriage. "He was absolutely devoted to Piper," says Surratt. Despite the knowledge that his new love might have come to him with deep-seated issues, "Whatever he knew about her, Fred never said a bad word about Piper. At first we all thought that she was good for Fred."

At work, his colleagues noticed that Fred seemed softer, less the unyielding scholar and more human. It was thought that Piper had smoothed his rough edges, making him more understanding. Always, Fred had been a quiet man, but now he seemed genuinely happy, delighted in the marriage and in Piper. On Friday afternoons, when the work week drew to an end, he'd entertain Surratt and Deanna Matthews, who also worked in the faculty offices, with stories of Piper's menagerie. When the pet rabbit she gave him went missing, Fred looked sincerely worried. "When they found that rabbit, it was big news," says Matthews with a soft chuckle. "It was like the return of the prodigal son."

Yet, in hindsight, Surratt and Matthews would grow to believe the marriage was based on one concept: that Fred took care of Piper, sometimes at high cost to himself.

From the beginning, Piper struggled at St. Mary's. After they married, Fred paid her $10,000 tuition, telling friends he viewed it as a good investment in their future together. When her grades slipped to the point where she was on academic probation, he put the Harper's Ferry house up for rent and moved eighty miles away to San Antonio. For more than two years, Fred drove three hours a day to and from work, so Piper would have more time to study.

His friends never heard Fred complain. Instead, in typical Fred fashion, he used the time in the car to work. Before long he'd covered the dashboard, passenger seat, and console of his beloved Corolla with scribbled on yellow Post-it notes, hastily saved ideas, concepts, and plans that occurred to him during his drives. Despite the hours in the car, Fred never lost his focus. That year, with Professor Linda Putnam of Texas A&M University and two other scholars, Fred began work on a comprehensive text, the first handbook of organizational communication.

At UT, Fred thrived. Each semester, he oversaw two students, one working on a Ph.D., the other a master's degree, and taught an upper-level class or two, while he worked on his research. Vernon Miller, one of his Ph.D. students, would remember him as a mentor and a friend. "He taught us how to express our ideas, to write, and he truly became interested in us as people," says Miller. "Yet he was incredibly involved in his own work. We students had a signal. If we stopped at Fred's door and he looked up, he was ready to talk. If he kept writing, we kept walking."

When they did talk, Fred looked for the brilliance in Miller's ideas. "He was this really unassuming man," says Miller. "You'd never expect he was such an expert in the field. When he gleaned a good idea, his whole face lit up, and he had a radiance in his eyes."

By 1987, Fred's book with Putnam and the others, *The Handbook of Organizational Communication*, was released, garnering positive reviews and winning awards. UT rewarded him by making him a full professor and granting him tenure. Fred saw his contribution to the school as substantial, boasting that he'd taken the department from number two in the field to number one. Still, he didn't see himself as particularly brilliant. "Fred saw himself as a plugger, who hung in there and worked hard," says a friend. "And, he was right. Tenacity and determination were two words made for Fred Jablin."

Over the years, he'd show that same resolve when it came to his marriage, a determination to stay the course even through the roughest waters.

At St. Mary's, all didn't go as well for Piper as it did for Fred at UT. Her stint in law school would, in fact, be disappointing, but she did make it through. She graduated in spring 1986, but not at the top of her class, where a young lawyer needed to be to attract a lucrative position.

After she graduated, with no need to remain in San Antonio, Fred and Piper moved back into the Harper's Ferry house. Piper settled in to study for the bar exam, while Fred worked on his research. In addition, he cooked and cleaned, two things it would seem that Piper never quite had time for.

Yet, Fred didn't seem to care, and Piper could be truly charming. The Kuentzes still lived next door, and their children circulated in and out between the houses, visiting. When their daughter Tessa fell on a cactus, Piper retrieved a tweezers and spent an hour soothing the child, as she pulled needles from her skin. "Piper loved children, and she looked like she'd be the perfect mother," says Linda. With her menagerie of pets, one friend would look back and visualize Piper as the prototypical earth mother. "You could picture her with a breeze blowing her hair and wearing a crown of daisies, surrounded by children, birds, and animals."

Fred, too, loved children. In fact, John Daly would go so far as to say that Fred was fascinated by them. From early on, Fred appeared captivated by the Daly's oldest, Johnny, a lanky child Fred nicknamed "Stretch." When he visited, Fred doted on the boy, playing with him and taking the time so few adults do to actually sit down and talk to him. He and the child were so close that when the Dalys left for the hospital in the middle of the night to have their second baby, they asked Fred to stay with Johnny. The next morning, the toddler awoke to find his parents gone and Fred Jablin in the house. Rather than shrieking with fear, Johnny giggled and looked completely unconcerned.

When Piper didn't immediately find a job after law school, she painted watercolors of dreamy bucolic scenes and flowers. Fred found her an agent, the wife of one of his graduate students, and she began selling her paintings for $200 apiece.

The October after her graduation, Fred and Piper threw his annual Halloween party. There'd been fewer of them in all the disarray since Piper entered his life, but that fall their lives seemed to be settling into place. By then Piper had a job at the Hays County District Attorney's Office, thirty minutes from their home in Austin. She appeared to enjoy the job, where her main responsibilities were prosecuting traffic cases, mental commitments, and protective orders. At that year's Halloween party, Fred dressed as the playful wood nymph Pan, from Greek mythology, the horned overseer of goat herders and shepherds. In something of a reversal, the diminutive Piper dressed as a commando, with camouflaged hair and headband and her face streaked in black.

At the party, Piper talked of her exciting job, showing friends her jacket with "Hays County D.A." on the back and entertaining guests with stories of drug raids. Yet less than a year later she was unemployed. Later she'd say she quit because she didn't feel safe working as a prosecutor, claiming a man came to negotiate a plea bargain and began by taking out a gun and putting it on the table between them. Fred would describe her departure from the D.A.'s office differently, saying she'd been fired.

Whatever the reason, this time it didn't take as long to find work, and Piper secured a position as a staff lawyer at the Texas Association of School Boards, where she read cases and interpreted decisions for school boards throughout Texas. Since she didn't have to litigate, she didn't have the stress of the courtroom, and years later Fred would describe this as her best job.

In 1988, four years into their marriage, Piper and Fred began planning a family. Although most faculty wives kept their private lives out of the office, Piper had no such qualms.

In fact, on more than one occasion she called and breathlessly told Surratt or Matthews, "Tell Fred he has to come home now. I'm ovulating!"

In typical Piper fashion, when the big news presented itself, she announced the pregnancy with flair. That year, Fred went off to the annual International Communications Association conference in San Francisco. As he walked from a meeting surrounded by Daly and his colleagues, a man dressed as a stork in a diaper approached them, handed Fred a cigar, and announced Piper was pregnant. Everyone laughed, and Fred appeared delighted.

From that moment on he was enraptured with the prospect of fatherhood. He talked often of his excitement, detailing the pregnancy for all who would listen. And he planned, single-mindedly committed to building a secure future for his growing family. At times he was so methodical about it, colleagues found it funny. One day he approached Matthews with a pad of graph paper and a pen, saying he'd heard tales of childhood illnesses. "How much a year do I budget for ear infections?" he asked. He was serious, but Matthews chuckled and said she didn't know, that such expenses weren't predictable.

In hindsight, Fred had reason to be concerned; not only were he and Piper expecting a baby, but they were also building a new home, high up in the rugged pine and oak covered landscape west of Austin, in Westlake Hills. The Texas economy was down, and Fred, ever practical, had picked up a wooded one-acre lot in this affluent section of Austin for less than half of what it would have sold for a few years earlier. The building process, however, was a difficult one for Fred, who thought contractors should adhere to as rigid a schedule as he did and didn't understand why Piper had to continually change her mind about what she wanted. But the house turned out well, with dark tile floors and an art room where Piper could paint. Fred kept the house on

Harper's Ferry to rent out, calling it the children's college fund, and just weeks before the baby was due, they moved into the Westlake Hills house.

In July 1989, Jocelyn Rountree-Jablin entered the world, a beautiful child with a round, full face. Both Piper and Fred were delighted, so much so that Fred brought the video he took of the birth to the UT faculty offices to play for Daly and his other colleagues. Many were uncomfortable at viewing something so intimate, but Fred didn't appear to notice. He was happy with Piper, the new baby, and about being a father.

On Friday afternoons when the office quieted down, Matthews and Surratt listened to him talk about Piper and the baby, always in glowing terms. He was very much in love, and had never looked happier. Yet, they knew something he never mentioned: that his young wife had a formidable temper.

Like good British servants who observe but never speak of all they know, Surratt and Matthews had taken phone messages from Piper. While the other faculty spouses were scrupulously civil with them, Piper rarely minced words. Before and after Jocelyn's birth, she called the office often, sometimes ten times a day, more often than not furious. "You tell Fred I just fired the maid and he needs to come home right now," she'd scream, then hang up.

Nearly every time she called, Piper was demanding one thing or another, and more often than not the message began with an angry, "You tell Fred . . ."

Please call home, Matthews and Surratt wrote on pink telephone message slips they slipped in Fred's mailbox. At times, Matthews wrote *now* and underlined it, hoping to clue him in to what awaited him when he picked up the telephone. Among themselves, Matthews and Surratt talked. "Piper's on a rampage again," one would tell the other. Surratt began to wonder if Fred Jablin's young wife had psychological issues. It seemed odd to her that a man who was so supremely rational would have married a woman who could sound so incredibly irrational.

While Piper worked, Jocelyn was in day care provided by her mother's office. Fred dropped her off each morning and picked her up each afternoon, when he finished classes. He was as devoted to the new baby as he was to her mother.

Then Piper's work arrangement changed again. Little more than a year after assuming her position with the school board association, she quit, taking a more lucrative slot at Henslee, Groce, a law firm that specialized in representing school districts.

If the Jablins' marriage was at times rocky, Piper's life as a lawyer didn't go any more smoothly. She crowed to friends about how exciting it was working for a law firm with a big support staff, including runners and researchers, legal secretaries, even someone to get her lunch, but a year later, while she was pregnant for the second time, Piper was let go. She was furious. "She said that they fired her because she was pregnant, and she threatened to sue," says a friend. Fred would later say that Piper filed a complaint with the Texas Employment Commission regarding the firing, which was dropped when the law firm countered by pegging the grounds not as Piper's pregnancy but a lack of performance.

Piper didn't take the rejection well. Friends would later remember how furious she was, threatening to get back at a lawyer she claimed had plotted against her. "She said she'd get even. She was spitting-nails angry," remembers Chris Daly.

Years later Fred would say he never thought that Piper had to have the big job with the six-figure-plus annual paycheck, but had believed she would work and help make ends meet. The house in Westlake came with a high mortgage, and he had to hire household help, since Piper didn't consider cleaning among her responsibilities.

At the baby shower the faculty and staff threw, Surratt

and Matthews were surprised at how friendly Piper seemed, after they'd fielded so many angry telephone calls. John Daly's by then ex-wife, Chris, would remember Piper's second pregnancy as a difficult one, in which Piper exhibited wild mood swings. Once, Chris visited and found her sobbing. Sitting together on a couch, Chris comforted Piper while she cried, angry about her life. "Why am I with Fred, he's so boring," she said. "There are lots of men out there, and I deserve someone better."

Chris Daly didn't know what was wrong. "It just seemed serious and dark," she says. Other things would later come back to Chris, including how Piper talked incessantly of her family. She was proud of them, of all they did, especially Tina, who'd gone on to become a nurse practitioner and had opened her own clinic in Houston. "One minute she'd be talking about Tina as if she were a bit crazy, saying, 'You won't believe what my sister Tina did,'" says Chris. "The next minute, Piper was boasting about her and the whole family. She thought the world of them, as if they were all very special."

Daly knew that Fred didn't care for the Rountrees. Tina visited often, sometimes babysitting for Jocelyn. While she was there, Fred appeared on edge and talked of Piper's "whacky sister." In truth, he seemed less than fond of many in the Rountree family. He rarely attended their summer reunions on the beach in Padre Island, because he said they were all "a bit eccentric."

That March, 1992, Piper gave birth to a son, Paxton Rountree-Jablin. Like his sister, who was then three years old, he was a beautiful child. Afterward, Piper spent weeks in bed, without the stamina to do more than what was absolutely necessary to take care of the baby. It was a case of postpartum depression. "I didn't want to get out of bed," says Piper. "It got worse with each pregnancy."

Those feelings were at odds with the Piper who dubbed

herself "Momea" and enjoyed nothing more than being a mother. "Being a mother is my personality, my thing," she'd say years later. "It's my purpose in life. It's simply what I am at my core."

Paxton would become his mother's favorite child, and she'd call him her protector and her strength. A few months after his birth, Chris Daly saw Piper at a party and approached her, anxious to ask how she was. "I'm fine," Piper said resolutely. "We're all fine."

Yet, in truth, all wasn't particularly well. It was around that time that Fred would later say that he and Piper went for marriage counseling. Perhaps it was spurred by something that happened shortly after Paxton's birth.

Piper withdrew all her funds out of her Hays County retirement account, $1,605.05, and did something she'd been talking about since college. She added another $1,400 and paid $3,000 for breast implant surgery. Fred found out two days before the surgery, and he didn't want her to do it. He argued against it, worried about reports of leaking implants and problems breast feeding.

The year after Paxton's birth, Piper invited Lora Maldonado, the coordinator of the graduate program Fred supervised, and her children for dinner. Maldonado would never forget the evening at the Jablins' Westlake house. Piper drank wine and talked, laughing and high-spirited, while Fred quietly cared for the children and cooked dinner. As friendly as Piper was, Maldonado had a hard time separating her from the angry voice she often heard on the telephone calling for Fred. In Maldonado's view, Fred Jablin appeared to consider his wife too good for him, but in truth she thought he was too good for Piper. "In Fred's mind, everything was about Piper. He waited on her hand and foot," says Maldonado. "Maybe that's what he had to do to keep the family together, because Piper was the kind of woman who was going to do what she wanted to do, and nobody was going to stand in her way."

At work, Maldonado and Fred had, at times, been at odds. He demanded that everything be precise, and she simply wasn't wired that way. Yet he never raised his voice with her. "He'd get this look on his face, like, *sigh*," she'd later recall. "And then he'd explain how he wanted things done. Rather than get angry, he'd try to prove he was right, win you over to his point of view."

In 1992, not long after Paxton's birth, Piper took a part-time position that turned into full-time work with the Texas Classroom Teacher's Association, in offices in a graceful Victorian mansion on Guadalupe Street, just outside downtown Austin. Her job was to work with teachers, providing legal counseling and employment representation. Jeri Stone, the organization's executive director, found the new staff attorney perplexing. At times, Stone walked into Piper's office, adjacent to the reception area, and found the Jablins' large dog curled up at Piper's feet. Although they had a laid-back office, no one had brought a pet to work before. At other times, Piper showed up with one of the children, most often Paxton. One day Stone walked in on Piper in the lunchroom while she was feeding the toddler. "Piper was shoveling food in his mouth so fast he barely had time to swallow," says Stone. "She seemed just frazzled, not unusual for a working mom, but more so, to the extreme."

A year after Piper began, Stone called her into her office. "Piper's leaving was a mutual decision," says Stone. "For our part, it was that she didn't have a passion for her work. She didn't care deeply enough about the teachers and their needs."

Piper then tried to open her own, small law office, but that didn't go well, either. She had few clients, and Fred later complained that he spent more money on her expenses than she made. And then Piper was pregnant again.

Friends would remember this as the Jablins' darkest hour, not because they didn't want the baby, but because there were problems with the pregnancy. Well into the pregnancy something happened, and the baby was lost. At the time, Piper and Fred said that she had miscarried. Later, Piper would describe it as a therapeutic abortion, after tests determined that the fetus had a genetic abnormality. "They were both absolutely crushed, devastated," says Matthews. "It was a really hard time for Fred and Piper. Fred was incredibly sad."

Afterward, Piper sank into yet another deep depression, but Fred said nothing to his colleagues or friends, except that she was having difficulty dealing with losing the baby so late in a pregnancy.

By 1993, the year they lost the baby, Fred told Daly that he'd come to realize that working was perhaps not Piper's strong suit. She'd had five jobs in her six years since law school and made a success of none of them. Fred, always the pragmatist, told his friend that he had a quandary. "Piper wanted to stay home, and Fred couldn't afford that on his UT salary with their cost of living," says Daly. "He was a little miffed that Piper wanted all her creature comforts and wasn't willing to work. It didn't seem to Fred that she couldn't work, but that she didn't want to. It bothered Fred."

For a while Fred solved the problem by taking on consulting work, something he'd in the past avoided to concentrate on teaching and research. That ended when Piper complained he was gone too much and that she wanted him home more with her and the children. That left Fred with even more of a problem. The University of Texas was going through a lean time, and raises weren't in the offing. While his future was potentially bright there, Fred's attachment to his work, as strong as it was, came in second to his commitment to his family. Convinced he had to make a change, he began circulating his résumé and networking to find a new, more lucrative position. "It wasn't the money, it was what the money could do for the family," says Daly. "Fred knew

Piper wasn't going to work. He had children and college to save for, a roof to keep over their heads. To Fred it made sense. His mother had stayed home while he and his brother were young, and that's what Piper said she wanted to do, so he decided to do what he had to do to make that possible."

By then, in 1994, when the Jablins had been married eleven years, Daly noticed a change in his old friend. As Piper became more extreme, more flighty and disorganized, constantly throwing their lives into turmoil, Daly thought Fred reined himself further in, becoming even more methodical and organized. It was as if to make the marriage work, he was becoming even more of Piper's opposite.

Fred was presented with an opportunity that year, a tenured slot in the Jepson School of Leadership Studies at the University of Richmond, in Virginia, at nearly double his UT salary. The position had its good and bad points. On the downside, he would be teaching undergrads, no longer grooming graduate students to be his protégés. But the school was new, and the first of its kind in the U.S. to grant a degree. Like getting in on the ground floor of organizational communication, that newness offered a certain allure. Fred enjoyed being a pioneer, and this was a second chance to be in at the inception of something exciting.

Later, he would also say that he saw the move as a way to give his family, especially Piper, a fresh start, in a new state, a new city, with new people. It appealed to him that they'd be moving farther away from her family, who in his view brought stress into the marriage. At the University of Richmond, he could make enough to support his family without relying on Piper for help, something he now viewed as a necessity.

Fred took the offer. Home values in Austin had climbed, and he and Piper sold the Westlake Hills house for more than they had paid for it, reaping enough profit to pay up all their debts, including credit card bills. Then they hired a moving van and headed for Virginia. "When they left, we all thought they'd make it," says Margaret Surratt. "In many ways, they were a really good couple, and they both seemed to love each other and want the marriage to work."

Years later, Tina would admit that for her the separation was painful. "I never forgave Fred for moving Piper so far away," she said. "What right did he have to take her and the children from her family?"

Not long after, Chris Daly ran into Piper in Padre Island, Texas, where she was at the Rountree family reunion, frolicking with Paxton and Jocelyn on the beach. Fred hadn't come. Chris would always remember watching Piper in the water. "Piper looked so happy, so excited just to be with her children," she recalls. "I thought that everything must be working out for them, that the move had been good for them. But I never forgot that day when she was pregnant with Paxton, when she sobbed and said she didn't want to be with Fred. She'd seemed so dark and sad."

5

"Virginia is a conservative state," a refined man in a blazer and khakis explained one late spring afternoon, sipping iced tea at a round street-side table outside a small café in Richmond's historic Fan District. It was a peaceful setting, one of gracious old homes and storefronts. Not far away, the state capitol building designed by Thomas Jefferson was draped under canvas, in the midst of a restoration, and tourists prowled the grounds of St. John's Church, inspecting the gravestones and standing where legend had it that in 1775 Patrick Henry passionately uttered, "Give me liberty or give me death."

The Powhatan Indians once called this hilly, forested land Manastoh. A deep sense of not just Revolutionary War but Civil War history permeates Richmond, a city that grew outward from the banks of the James River. Richmond, after all, was the capital of the Confederacy, proudly attested to by Monument Avenue, a statue-laden boulevard commemorating Southern heroes Robert E. Lee, Stonewall Jackson, J.E.B. Stuart, and Jefferson Davis. When a likeness of African-American tennis great Arthur Ashe, a Richmond native, was added at the intersection of Monument and Roseneath Road in 1996, it caused a furor from both those who argued its presence violated the intent of the landmark and those who said it cheapened Ashe's memory to honor him among men whose proslavery ideologies had torn the nation apart.

"By conservative, I mean there are things we just don't talk about here, things it would be impolite to bring up in conversation, like race and infidelity," the man went on to explain. "But please, don't confuse that. Not talking about something doesn't mean it doesn't exist. If a man or woman has a dalliance, it's of little consequence. More important is to keep up appearances. We have a sense of Southern propriety in Virginia. How one presents oneself is uppermost."

Those who travel northwest through Richmond eventually enter Henrico County, a half-moon crescent surrounding the northern half of the city, dead-ending on both sides at the James River. In Henrico, the population is predominantly white and prosperous. In this affluent setting, the West End—a patchwork of subdivisions cut out of the thick forest—is among the wealthiest and quietest. It was here, in 1994, that Professor Fred Jablin and his wife of eleven years, attorney Piper Rountree, purchased a red-brick and wood-sided two-story house with shutters, set back from the street, in the thirty-year-old Kingsley subdivision, at 1515 Hearthglow Lane.

Richmond's West End was a calm place to live, one without the crush of the city. On weekday mornings, fathers and mothers drove off to work, after watching their children board school buses. At the nearby Gayton Crossing shopping center, West Enders shopped for groceries at Kroger or the upscale Ukrop's, and wine, bakery items, clothing, toys, or gadgets at small specialty stores. In the local Starbucks, customers queued up for morning lattes, half-cafs, and mochas and talked neighborhood news, their children's soccer teams, and school events. Occasionally the subject turned to more private matters, including what they saw or heard happening at the houses next door.

At first glance the Jablins with their two young children—Jocelyn, five, and Paxton, two—fit in well. From the outside, they were a picture perfect family, with a house full of book-cases, couches, lamps, and all the latest appliances. They installed a dark oak table with a sideboard in the narrow dining room, and Piper turned the bonus room—a converted, attached one-car garage—into a craft and sewing room with an aquarium. Fred claimed a room built into the detached garage as an office/workshop. In the large, treed backyard, Fred planted a cutting from his grandmother's favorite rose-bush, as he had at each of his homes. With his Ford Explorer and her brand new Chrysler Voyager minivan in the drive-way, they appeared a happy family. Perhaps, at least in the beginning, they were just that.

"At first, they seemed like a really good couple," says a neighbor. "Fred was smart and successful, excited about his work. The two kids were darling, nice kind of quiet kids. But then there was that Piper thing. From the beginning, she never fit in."

"Piper was this petite woman with blond highlighted hair. The rest of the moms wore khakis and loafers, but she wore skintight jeans and cowboy boots," says another neighbor. "She was out there, flamboyant. Everyone noticed her."

Some neighbors found the difference in Piper a welcome change. "My husband said, 'At least she's not just another typical West End Southern belle,'" says a neighbor. "I wasn't insulted. I knew what he meant. Piper, well, Piper was . . . different."

In the mornings, Fred drove off to work at the University of Richmond, a fifteen-minute trek through the woods on winding residential streets. The campus was breathtaking. Founded by Baptists in 1830 as a men's seminary, it had been moved to the shores of Westhampton Lake just after the turn of the next century, on the site of an abandoned amusement park. With 360 acres of forest and rolling hills, the university hired Ralph Adams Cram, an architect who'd designed major buildings at

44 / Kathryn Casey

Princeton and West Point, and the result was a school that exemplified Cram's "collegiate gothic" style.

At times the school had been in financial turmoil, nearly closing or selling off to the state of Virginia, but UR's alumni had always come through, as in 1969 when E. Claiborne Robins, a 1931 graduate whose family's pharmaceutical company marketed the cough medicine Robitussin, gifted the school with what was then a staggering sum: $50 million. Ironically, A.H. Robins pharmaceutical, which also manufactured the Dalkon Shield, would later fall victim to a decade of litigation and eventual bankruptcy. By then, fueled by the gifts of Robins and others, UR had one of the largest endowments in the nation: in 1994, when Fred Jablin arrived, approaching $1 billion. Most of the students came from East Coast families, and the costs of attending were $30,000 per student per year. For qualified students who couldn't pay, the university had ample funds for grants and scholarships.

The Jepson School, where Fred signed on, began under similar circumstances, funded by a $20 million gift from Bob Jepson, an alumnus and corporate turnaround specialist, who wanted to start a leadership school and gave the university the funds to realize his dream. The Jepson School opened in 1992, the year the University of Richmond hosted presidential debates pitting the first George Bush against Ross Perot and a charismatic Arkansas governor, William Jefferson Clinton.

The Jepson faculty had impressive credentials. In Jepson Hall, with its arched bell tower, Dr. Joanne Ciulla, one of the school's founding professors, had an office on the first floor, directly next to Fred. She had a Ph.D. in philosophy from Temple University and had come to UR from the Harvard Business School and the Wharton School. At UR, she taught ethics and leadership. Other faculty examined leadership through the eyes of their diverse disciplines: psychology, religion, government, art, history, and political science.

For Fred, the Jepson slot had presented itself at just the right time. He'd had other offers that year, but nothing as lucrative. And it held another appeal: Despite Fred's reputation and accomplishments, there were other top-drawer communications scholars at UT, like John Daly. At UR, Fred was offered what might have taken him a decade or more at UT: an endowed position, the E. Claiborne Robins Chair, a lifelong, guaranteed post, with ample time to do his research. "At the University of Richmond, Fred was a big fish in a small pond," says Knapp. "That held a certain appeal."

The benefits weren't one-sided. The Jepson School had reasons for courting Fred Jablin. "We wanted Fred because of what he'd done for organizational communication, research that used meticulous standards," says Ciulla. "Leadership was also a new field, and it needed that discipline."

The school's mandate was to study leadership and to teach students what they needed to know to lead. Students applied as juniors, and the competition was tough, with 150 vying for sixty openings.

At first, Fred's adjustment to working with undergraduates was a difficult one. But before long he settled in and enjoyed the change, teaching classes in not only communication and leadership, but collaborating with other faculty members to develop classes in "Art and Leadership" and "History and Theories of Leadership."

In many ways the Jepson School was a good match for Fred Jablin. He was a man who believed in rules, and it was an institution that studied morals and ethics. Quickly, his office, Jepson 242, filled with framed drawings by Paxton and Jocelyn, and photos of Piper and the children. Shelves strained under his collection of books, and stacks of papers covered his desk, shelves, file cabinets, and even the floor. In the spring, pink and white azaleas bloomed on the campus, and in the fall the leaves turned the surrounding forest a brilliant scarlet and gold. As the years in Virginia passed, Fred's office and the University of Richmond must have seemed a logical place to him, a refuge where people treated each other with respect.

All was not as tranquil at the Jablin household at 1515 Hearthglow Lane.

"The West End is a beautiful place," says one of the Jablins' neighbors. "You look around at the nice houses and you think people must be happy. But sometimes, they aren't."

In the beginning, Piper and Fred appeared, at least to neighbors, to slip naturally into their respective roles. In the mornings, Fred drove off to work and Piper made fresh bread for the family and even baked homemade biscuits for her dog. While he was at the university, her days were spent playing tennis, at the children's activities, and volunteering at their schools, especially when it came to art class. Throughout their years in Virginia, Piper would tutor her own children, working with them with their art, giving lessons to Jocelyn's Brownie scout troop. "Every year the Jablin kids would get a ribbon at the school art competition," says one neighborhood mom.

In the evenings, Fred threw a ball around with Paxton in the front yard and then cooked them all dinner. He built a playground with swings for the kids, and Piper quickly filled the house and yard with her menagerie of animals, including dogs, cats, rabbits, birds, and ferrets she walked on a leash. She took in wounded animals and nursed them back to health, even building a small pond in the backyard that she populated with fish and fist-sized bullfrogs she'd found at a nearby creek. In no time, Piper's bullfrogs, to the dismay of the neighbors, had spread out and taken over the street.

On weekends, Piper and Fred worked in the yard together. She planted azaleas, and Fred constructed a deck off the kitchen and a lattice gazebo. One year, for Mother's Day, Fred built Piper a raised bed where they grew vegetables, including peppers she cooked with tomatoes and turned into homemade salsa. Piper was proud of her talents in the garden, and when one neighbor remarked that she'd planted a wisteria two years earlier and it had yet to bloom, Piper boasted that she'd put hers in the ground just that summer, and it was covered with blossoms. "She was always doing the oneupsmanship thing," says the neighbor. "You could never do anything as well as Piper, not to her mind."

Yet, at the same time Piper was kneading bread every morning, the real work of the household—as in Texas—was done by a parade of nannies who came and went, many lasting only days before they quit or Piper fired them. Most were in the U.S. illegally and spoke little English, and along with watching over the children they were charged with caring for the house.

"It seemed like the only criteria Piper had for hiring them was that they had a pulse," says Melody Foster, an ample, warm-hearted woman with shoulder-length dark brown hair and a ready laugh, who with her husband, Pete, lived directly behind the Jablins. As they had with the Kuentzes, Fred and Piper became friendly with the Fosters, who had a daughter, Chelsea, Paxton's age; so much so that Fred cut a gate through the back fence so they could circulate freely between yards. Along with both having young children to care for, Melody and Piper had something else in common: They were both attorneys.

A year after moving to Richmond, Piper became pregnant again. This time it was an ectopic pregnancy, and she again lost the baby. Afterward she returned to her very dark place. "It's like everything goes gray," she says, explaining what her postpartum depression felt like. "Everything feels dull."

On weekends, Mel and Piper sometimes opened a bottle

of wine and sat together and talked while they watched the children play. At such times, Piper often brought up her family in Texas and how much she missed them. Each summer, she took the children to visit, and when she returned she seemed melancholy, yearning for her family and talking often and long about all the Rountrees, especially her closest sister, Tina. Piper wanted to be with them, Mel understood, but it seemed almost unnaturally so.

Family was important to Fred, too. At work, many were noticing slight changes in him the year after he arrived at UR, when both his parents died within a six-week period. Stunned with so much loss in such a short period of time, he became introspective. He spoke with Joanne Ciulla about life and death, and the loss of a loved one. "He seemed to be grappling with some very deep feelings," she says. "He loved his parents very much, and he was hurting."

It would seem that he got little comfort at home. Piper didn't go to either of the funerals. It had been years since she'd attended family occasions at their house. After they'd fallen ill, Fred had driven to stay with them over weekends, to help. Piper never went along. Still, when his $130,000 share of their estate arrived, Fred invested it in something he and Piper both wanted, a beach house they dubbed "The Outer Banks," four hours away in Corolla, North Carolina. It wasn't fancy, and over the years he'd spend weekends and vacations rebuilding the small house, but it was a place to escape to.

That December, Piper gave birth for a third time. She and Fred named the baby Callyn Rountree-Jablin and called her Callie. Afterward, Piper again sank into an even deeper depression. When it passed, this time motherhood didn't come as naturally for Piper.

Callie, a golden child with blond curls and crystal blue eyes, was a different kind of child than Jocelyn and Paxton. Joce was a happy yet easy young girl, by then six years old, and she demanded little. Three-year-old Paxton loved sports and was all boy, yet never seemed to push his mother for what he wanted. On the other hand, neighbors would describe Callie as an impish child, one with a mischievous glint in her eyes. While the two older children, even from young ages, tended to take care of themselves, Callie, by her very nature, demanded attention. "She was the type of kid who needed her mom more," says Mel. "And Piper didn't appear to have more to give her."

By the time Callie had turned three, Piper was escaping to the Fosters' house, "just to get away for a little while." Mel would look outside and see Callie in the yard, by herself. When she voiced concern, Piper assured her, "Oh, I do it all the time. She's all right."

For a long time it seemed odd to Melody that a woman like Piper, who appeared to base her entire self-image on being an exceptional mother, could be so careless with her children. At times Piper appeared the ideal mother. She became interested in geology and spent hours with the three children at the creek, picking up rocks and identifying them. She painted castles and Walt Disney princesses on the walls of the girls' bedrooms, covered the walls of the downstairs bathroom with a mural of trees, and painted the doghouse to look as if it were built of red brick. When she volunteered as a guest reader in one of the children's classes, she arrived dressed as Alice in Wonderland, that day's book. At other times, however, when Piper was in one of her funks, she locked the children out of the house and simply went to bed.

From week to week nannies came and went at the Jablin household, some quitting, others fired. When she was between help, Piper became anxious, asking Mel or, on days Mel worked, Mel's nanny to babysit. Often Piper seemed to be searching for a way to absolve herself of the responsibilities of the children, like at bunco one night. While Piper played the popular dice game with the other neighborhood women, she proposed starting a babysitting co-op. Some of the women agreed it sounded like a good idea, but no plans were made. Still, the following morning Piper showed up with Callie in tow, trying to drop her off at the home of one of the women. Piper gained a reputation in the neighborhood, and some of the women began checking their doors when the doorbell rang. If she stood outside, they wouldn't answer. "She was constantly trying to find others to care for her kids," says Mel. "She loved playing with her children, teaching them art, whatever her project of the week was. But she just didn't want to take care of them."

Yet at first the two women got along well. Mel found Piper creative and fun, if eccentric. Kingsley was a friendly place, the kind that fosters neighborhood parties, bonfires, and barbecues. At one Halloween party, Fred wore his favorite wizard costume and Piper came attired in a full fencing outfit, with a sword. More often than not Piper had a glass of wine in her hand. She never made a secret of her drinking. When Mel's father came to visit, Piper walked in the Fosters' back door unannounced, as she often did, and introduced herself to the elderly gentleman as, "Piper, the neighborhood drunk."

In the summers, most of Piper's time was spent on the tennis courts. The family joined Canterbury, a neighborhood tennis and swim club, for a cost of $300 a year. One summer Piper caused a stir at the club swimming pool. She still ran every morning and had the trim figure of a teenager, but members complained that they didn't want a parent wearing a thong bathing suit in front of their children.

A while later, without consulting Fred, Piper also joined Raintree, where the dues were $135 a month. Unlike Canterbury, Raintree had day care, and Piper dropped the children off and left them there much of the day, far beyond the club rule of a three-hour maximum. At such times, although parents were supposed to be on the club grounds, Piper was nowhere to be found. It was then that the rumors began: that Piper was having affairs with the young men she played tennis with at the clubs. Melody heard the talk, but never said anything to Fred. "It wasn't the type of thing you told a spouse," she says. "It was gossip."

Through it all, the Jablins' marriage remained troubled. In 1996, the year after Callie was born, they were again in marriage counseling. Piper was unhappy, feeling unappreciated and ignored. She expected more from Fred, more attention to her and the children. And there was the thing about the money. During the sessions, Piper complained that Fred kept her on a tight financial leash, and said she wanted to be the one to control the family funds. Fred agreed. From that point on he set aside $800 a month to cover large expenses, like insurance, and gave the balance of the money to Piper to pay the bills. "Fred wanted her to have a sphere of influence," remembers John Daly. "She needed something to be in charge of."

When Daly saw Fred at conferences after the move to UR, Fred seemed satisfied with his decision to leave UT, yet there was a sadness about him. Fred told his old friend that he worried about Piper. She wasn't content in Virginia, he said, despite not having to work and being free to spend time with the children. "Piper could never seem to be happy," says Daly. "That made Fred feel bad, that no matter what he did for her, it wasn't enough. He believed she was troubled, but he loved her. So he had no choice but to keep trying."

Early during her time in Richmond, Piper connected with a kindred soul, Loni Elwell, another well-educated, West End, stay-at-home mom. Loni, married to a corporate executive, had a master's degree in social work. She'd landed in Richmond much as Piper did, because her husband's job brought her there. Their daughters were on the same soccer team, and both their husbands worked long hours. "We met at a soccer game and became fast friends," recalls Loni, an athletic woman with short blond hair and a wide smile. "We'd hook up in the morning and spend all day together. We were pretty much inseparable from the start. Everyone walks and talks the same in the West End, drives the same kind of cars. Piper and I were different. True and honest."

"They considered themselves non-Stepford wives," says someone who knew them both. "They kind of laughed about the other women on the West End."

By then, Melody Foster had become a stay-at-home mom, and she was growing weary of Piper's drop-ins, simply walking unannounced in her unlocked back door. Once, when the school nurse called and Melody's youngest daughter, Claire, was sick, Piper, who was in Mel's kitchen drinking coffee and rambling on, said, "Oh, just don't go get her. That's what I do."

It bothered Mel, too, that Piper took things to the extreme. The year the Fosters cut down a few trees in their backyard to make room for a playground for their children, Piper protested by sitting on one of the fallen tree trunks. And then there were the constant complaints. Piper didn't like a dusk-to-dawn light the Fosters installed in their backyard, claiming it made a buzzing that kept her awake at night.

Eventually, Melody Foster concluded that despite Piper's demonstrative love for her children and her view of herself as the ideal mother, Piper didn't seem to understand her children's needs except through the distorted glass of her own wants. "Piper would get wound up and go on binges, like rollerblading," she says. "She expected the kids to rollerblade with her constantly, every day, whether they wanted to or not, until her obsession with it passed. She loved tennis, so they were expected to love tennis. A lot of what Piper did just seemed exaggerated and strange."

After a while Melody began locking her doors and pulling her car into the garage, to make it appear she wasn't home, in order to avoid Piper planting herself in the kitchen, wanting to talk and not taking the hints when Melody was ready for her to leave.

In contrast, Loni Elwell had no such qualms. She and Piper walked unannounced into each other's homes. "Peeps, where are you?" Loni would call out. Then they'd spend the day together, playing tennis or doing things with their children, and often cooked dinner at one or the other's house at night. "My husband was gone a lot and Fred was always working," says Loni. "It was the way our lives were then. We took care of the children and our husbands worked."

Later, Loni would describe Piper as misunderstood, a kind soul who many simply didn't appreciate. "Piper lived in the moment," she'd say. "She was the one who taught me to wear my good clothes and drink my wine out of my best crystal, not saving it for a day that might never come."

Yet, Loni experienced Piper's gray side, the depression that seemed to haunt her life. At times, she'd arrive and find Piper in bed in the middle of the day, with all the blinds pulled. "Peeps, what are you doing?" she'd prod.

Piper would just want to be left alone.

Over the years, Loni thought she could see the breakdown in the Jablins' marriage. But while others talked of Piper's histrionics and her unreliability, Loni blamed Fred. He worked long hours and was rarely home. When he was, he spent the time in his home office. Mel saw that, too, a man devoted to his work, who knew his wife was troubled and propped her up by hiring nannies and cleaning ladies, by sending Callie, the youngest, to day care, where he knew she'd be cared for, because he didn't trust Piper to watch over her. It must have been disappointing for Fred. He'd left UT to give Piper the opportunity to stay home with the children, but over the years in Richmond, it became apparent that she didn't manage the family any better than she had her career.

There were other indications as 1998 drew to a close that all wasn't well. By then Jocelyn was nine, Paxton six, and Callie three, but those who knew the Jablins sensed a distance in their relationship. "There was a coldness that kind of settled into the marriage," remembers Loni. "You rarely saw Fred touch Piper, even on the shoulder."

Of all the children, Jocelyn seemed to be the most disturbed by the family troubles. Over the years growing up on Hearthglow, the Jablins' oldest had blossomed into a vivacious and creative child, spunky and full of life. She led the other neighborhood children when they put on backyard plays, and spent hours laughing in the Fosters' playroom, directing the younger children as they played games. Yet, as Piper deteriorated, she put more and more responsibility on Jocelyn, who was forced to take over the tasks her mother left undone. Some nights, Joce walked outside to the deck, where her mother sat drinking wine, to ask what frozen entree she should put in the oven for dinner.

A doctor who treated Piper saw something else in her. He got the impression that she was on the lookout for a new man in her life. When Piper had an appointment, she flirted with him, until he issued a standing order: From that point on, a female nurse was in the examination room with them at all times. "I had the feeling Piper would cross the line," says the physician. "I didn't want her to have that opportunity."

Meanwhile, in his classroom at the University of Richmond, Fred Jablin taught his students many concepts, including one he called "the Dance," an analysis of the way people interact, the give and take between two people in a relationship, business or personal. Many had the feeling that all was not well in the Jablin marriage, that Fred and Piper's "dance" was winding tighter and tighter. When Joanne Ciulla and her husband went to dinner at the Jablins', his strong reaction to Piper surprised her.

"I hope we don't have to do anything else with them," Ciulla's husband confided in her later. "That woman is frightening."

6

. . . .

I'm a very complex person," Piper told Dr. Steven Welton, a psychiatrist at Richmond's Institute for Family Psychiatry, on January 15, 1999. It was her first visit in what would become yet another cycle of therapy. The next two years in the Jablin marriage would be documented in the notes Welton wrote about their sessions together, painting a portrait of a woman consumed by her own wants and needs.

While Piper talked, Dr. Welton listened, interjecting questions. That first day, Piper talked of her childhood, saying her early years had been tumultuous and tainted by a family history of alcoholism. When asked to describe herself, she said she had a high IQ, 138, and a competitive, type-A personality. She said she drank wine daily, and told the psychiatrist that during college she had been diagnosed with attention deficit disorder. If she disclosed her other problem—the bulimia Fred would claim she had—Welton didn't note it on her chart. Piper talked about her depression, her bouts with despair after the births of her children, and the blues that sent her to bed for days at a time, especially during the Christmas holidays, when what should have been a jolly season propelled her into melancholy.

When it came to her marriage, Piper described it as faltering. She saw Fred as uninvolved and disinterested and made no secret of what many had suspected in Virginia: that she'd been unfaithful to him. In fact at the time she began seeing Welton, Piper said she was involved in an affair with a younger man, one just in his twenties. Through it all, she remained so upbeat, Welton attempted to explain her demeanor in his records: "There is an impression that her glib surface adaptability hides some clear struggles with sense of self, identity, and relationship problems." Defining his goals for Piper, in addition to her current diagnoses of depression and attention deficit disorder, Welton noted that he'd consider alcohol abuse and a mood disorder. That first session, he took her off the Ionamine for her ADD and put her on a more common drug for the disorder: 20 mg of Adderall.

At first, changing her medication appeared to work. At their next meeting, seven days later, Piper said the Adderall had been helpful, increasing her ability to focus, especially when playing tennis. By then she was also on the antidepressant Prozac, Welton noted, "as she had been more frantic and stressed than usual." The physician set a follow-up for two weeks and noted his goal on her chart as determining her best combination of medications.

It was about that time that Melody had what she'd describe as an even stranger than usual conversation with her neighbor. One morning when Piper dropped in for coffee, Melody was griping about her husband, Pete, nothing serious, just the minor annoyances of everyday married life. Piper chimed in, complaining about Fred. Then Piper announced that *she'd* found a way to eliminate such aggravations. "You should put Pete on Prozac. I've got Fred taking it, because I think he's depressed," she said. "He doesn't know. I've been slipping it in his coffee."

Years later Piper would contradict Melody Foster's recollection, claiming that Fred had been prescribed the antidepressant by Tina's second husband, Howard Praver, a Houston gynecologist. Praver, however, would deny that he ever treated

Fred as a patient and say he never wrote a prescription for him. "Fred didn't need Prozac. He was one of the most level-headed, laid-back people I'd ever met," he says. "I had no reason to treat him for depression."

At the time Piper told Melody about the Prozac-laced coffee, Mel was stunned, not knowing what to say, wondering if Piper had just made it up to shock her. Or perhaps, she thought, it was a product of what she increasingly viewed as Piper's disturbed mind. Mel quickly changed the subject, thinking there were some things she'd rather not know.

Piper's therapy with Welton continued throughout that winter and spring. At times she talked about a relationship with yet a third man—beyond Fred and her twenty-something lover—this time a nonsexual liaison with a man she described as "a match for her intellectually."

In March, Piper missed all her appointments with the psychiatrist, and didn't show up until early April. She seemed sanguine then, saying her mood had been stable. During the session, she claimed she'd cut off the relationship with her young lover when she discovered he was having an affair with one of her friends. Perhaps having the younger man out of her life forced her to focus, at least briefly, on her marriage. Piper is "now debating about the value of her relationship with her husband," Welton wrote. "In general, she feels stuck with inertia on her part. Suggest need her marital Rx to work on issues with husband."

In hindsight it seems that summers were usually more tranquil times in the Jablin household. Piper enjoyed her trips home to the Rountree family reunions on the Texas beach. Fred still rarely went, but he didn't interfere with Piper attending and taking the children. And in summer, without classes to teach, he was able to spend more time with her. Piper seemed to require that: Fred's constant attention. She only seemed happy with the marriage when he focused on her and the family. As soon as his thoughts strayed back to work, Fred would later say, Piper grew discontent.

The summer of 1999 started out well. They spent time at the beach house and the neighborhood pool. But their usual hiatus of marital woes was interrupted by a discovery: When Fred decided to buy a car, a new Ford Explorer, the finance company ran a credit check. In the process, Fred discovered that during the three years Piper had overseen the family finances, she'd racked up $32,000 in credit card debt. Fred was furious, and worried. "It's just more proof of Piper's instability," he told one friend. "I can't believe she's doing these things."

A careful man, Fred couldn't tolerate owing so much money. After coming to grips with the initial shock, he transferred all the debt onto his own credit cards. His plan was to take charge of the family finances and to begin paying the debt off. To ensure that it didn't happen again, he took away Piper's credit cards, leaving her with only one that carried a $500 maximum. Finally, he announced an end to his wife's way of life. No longer would Piper be free to spend her days as she pleased, playing tennis while he worked and others cared for the children and the house. Fred told Piper that he expected her to work, to help pay off the bills.

It wasn't just lip service. Fred took action.

Determined that she'd find a position, he networked and quickly found a retiring Richmond attorney who needed someone to take over his practice. There was a problem, however: Piper had never been licensed to work as an attorney in Virginia. When she'd first arrived, she could have petitioned the state bar for reciprocity, basically asking Virginia to grant her a license based on her Texas license and experience. But she never applied, and that grace period lapsed. Unlicensed, Piper could legally work only under another lawyer's supervision. When she told Mel about the work she'd taken on, however, it sounded as if—Virginia law license or not—Piper was doing her own work. As usual, she wasn't letting details—this time even breaking the law—get in the way of what she wanted.

In early August, Piper didn't seem particularly worried. She told Dr. Welton that the summer had been a good one, and she seemed excited about working again. "She's struggling and trying to define some goals for herself," he wrote. The idea of setting long-term goals frightened her, Welton suggested, surmising that the central issue could be her reluctance to take the next step, to come to terms with the need for her to take her law boards to gain licensing.

In late September, Fred was back in the classroom teaching at UR, and the old Piper returned, a woman easily overwhelmed and sullen. She complained to Welton that four-year-old Callie didn't like her day care and was acting up, and that Fred wasn't devoting the time to her and the children that he had over his summer vacation. Piper referred to it as Fred "avoiding his responsibilities" at home.

"Very angry with husband and feels hopeless that he will take responsibility," Weldon wrote on her chart.

That fall, she continued to practice law, but after a client complained, the Virginia State Bar sent Piper a letter, questioning her lack of a license. Piper responded by filing a formal application for reciprocity, claiming she was still within the grace period because she'd continued to work for Texas clients while living in Virginia. The clients Piper listed were her family members, including Tina and her clinic. To bolster her claim, Piper asked Mel to sign a form stating that she was aware of Piper's continued legal career.

"But I don't have any knowledge of your doing any legal work since you moved to Virginia," Melody told her.

"It's only a formality," Piper assured her, giving her the form on which she'd already written what she wanted Mel to say.

After Piper left, Mel took the form and wrote across it: "I

was given this but I have no knowledge of what's stated here." She then signed and mailed it. Weeks later, when Piper tried to put an ad offering her services as an attorney in the Kingsley directory, Mel called the neighborhood association president and told her that Piper wasn't licensed. They didn't run the ad.

Mel wasn't surprised by Piper's actions. Years earlier, when Mel had complained about being bored at continuing education classes, required by the state, Piper told her to simply lie on the paper, sending in the forms saying she'd attended the sessions when she hadn't. "I told her it wasn't ethical and that I didn't want to risk losing my law license," says Mel. "Piper told me that's what she used to do, and that I'd never get caught."

It undoubtedly didn't help Piper's appeal for reciprocity that she showed up at the state licensing board office to turn in her paperwork wearing a gold bikini under a sheer cover-up. Not long after, her petition was denied. From that point on, the only way Piper Rountree could be licensed to practice law in Virginia was to pass the state's written bar exam.

By December 1999, Piper was on Adderall for her ADD, as well as antidepressants and an antianxiety drug, Xanax. In an attempt to ease her through what was becoming an increasingly difficult period, Fred paid $3,000 for a University of Richmond Law School review course to help her pass the Virginia bar exam. To free her from responsibilities while she studied, he hired a full-time housekeeper. For years he'd been propping her up, hiring help, pitching in to do what she seemed unwilling or unable to do. On her last session with Welton for the year, Piper seemed happy that Fred had taken over most of her duties with the children. Welton noted her progress toward her goals: "looks positive at this point."

Just after the first of the year, in January 2000, Piper began the UR review course. With the other students, she talked as if it were a formality, that with her experience, including time as a Texas prosecutor, she had no fears of failure. Instead of following the lectures and highlighting details to study, Piper sat in the back row of the auditorium with a sketchbook, drawing.

Outside the classroom, she concentrated not on preparing for the exam but continuing to enjoy her life. Even with the law boards lurking, Fred would later say he rarely saw her study. Despite everything he'd been through with her, however, Fred remained optimistic. For the week after the exam, he booked a three-day, spring-break trip for the two of them to the Bahamas. It was to be a celebration, to commemorate her passing. Perhaps he also saw the time together as an opportunity to rewind and get to know each other again, to mend the deep cracks developing in their increasingly tenuous relationship.

Despite his plans for a victory celebration, in March, when the results were posted, Piper had failed.

Although she'd spent little time studying, the failure plunged her into yet another deep depression. She didn't want to go to the Bahamas, and when the trip was cancelled due to a problem with the plane, Piper complained about Fred's alternate plan, to spend the weekend together at a posh hotel. Later he'd say she went reluctantly and stayed in the hotel bed throughout that weekend, disinterested in him.

In April 2000, Piper saw Welton for a session and told him she'd failed the exam. Absolving herself of any responsibility, she claimed to have studied hard for the test and recited a laundry list of reasons for the failure, everything from the recent death of one of her cats to an imbalance in her hormones. Piper even talked about the tumult of having one housekeeper quit, only to hire another who she claimed had stolen jewelry and her medications. That afternoon, Piper told Welton that she'd been on an emotional roller coaster, fighting off anxiety and depression. She was so upset one day that she went to an emergency room for help. After listening to her description of the events, Welton characterized the event as a severe panic attack.

Two weeks later, in mid-April, Piper's mood had spiraled even further. By then she was on Xanax, Adderall, Prozac, and other drugs. Painting herself as an unappreciated victim, she complained to her therapist that Fred didn't understand how fragile she was, and said she was furious he was planning a summer trip to Disney World with the children. While many mothers dream of such a vacation with their families, one filled with fantasy and fun, Piper saw it as beneath her. "I can't believe he wants to take the kids to such a common place," she told one friend. Piper saw Disney World as a haven for the "Stepford" wives she complained about in the neighborhood, the women who went to work and cared for their families, those who didn't live as she did, seizing the days for her own enjoyment.

It was in early May when Fred met with Welton for the first time. Piper had been talking about her husband in therapy for seventeen months, painting a picture of a man who minimized his wife. But in the psychiatrist's office, Fred appeared sincerely concerned about Piper's problems. In fact, he said that he worried she was so troubled she could commit suicide. "Husband seems to understand," Welton wrote. There was something that day, however, that did worry the psychiatrist: Piper was on Xanax for her anxiety, and Welton noted that she was using increasingly larger amounts. "Explore the potential for addiction," he noted.

Just weeks later, the Jablin family trip to Disney World was postponed, and on May 22, amid concerns that she had an ovarian cyst, Piper went in for a hysterectomy. Afterward, friends noticed that she looked drawn and dark, exhausted. Fred would later describe her condition as a "general malaise."

Soon it was summer again, and he was home to take over on the child care and household responsibilities. As Piper's physical condition slowly improved, Fred continued to care for the children, while she disappeared for large portions of the day, painting. During therapy sessions she complained to Dr. Welton that while Fred was supportive, he pushed her to pull her life back together. "Feels like she is grieving the last year, failing bar exam, losing cat, preoccupation that has kept her out of children's lives," Welton wrote. "Goal: Urged to allow self-healing time."

Later Fred would call this Piper's "earth mother period," during which she told him angels communicated with her. As a scientist, a man who saw the world in concrete terms, Fred must have struggled with a wife who communicated with invisible deities.

As the summer of 2000 drew to a close, Piper seemed uneasy with the marriage, complaining to Welton about a lack of intimacy. Yet by then the physician was questioning his patient's view of the world and her ability or willingness to make real changes. Twenty-one months into his sessions with Piper Rountree, the psychiatrist appeared frustrated, writing: "Every discussion seems to hit roadblocks . . . Piper remains unrealistic about money and seems to disengage from my active target of this . . . mood has digressed."

In September 2000, *The New Handbook of Organizational Communication: Advances in Theory and Research Methods,* Fred's follow-up with Linda Putnam to the first, highly successful handbook, came out in hardcover. His career, as usual, was rolling along, yet his marriage was becoming all the more troubled. In October, Piper complained to Welton that she was trying to control her spending and working on staying relaxed and focused. But he noted that the evidence was quite the opposite, that she was losing control: "[Piper] is using more Xanax . . . need to get a handle on this." Two weeks later things were no better. Piper told him she had trouble sleeping, and that Fred had moved into the spare bedroom.

That fall, Loni would later say, the strain in the Jablin household was so pervasive she could "feel" it. Piper dyed her blond-highlighted hair dark brown, her natural color, and more than one neighbor thought the change reflected her darkening mood. She looked ever thinner and more troubled.

In November, Jocelyn went into the hospital for minor surgery, performed by a Dr. Jim Gable. Later it seemed to Fred that the surgery marked a change in his marriage; from that point on Piper rebuffed his sexual advances.

Much of that fall, Piper continued to be missing from the household, ostensibly to paint. Fred was the one at the children's soccer games and at the school bus stop in the morning. As the holidays approached, he must have been worried. Piper already seemed fragile, and her most difficult time of year lay ahead. He made plans to drive the family to northern Virginia for Thanksgiving, to spend the holiday with his brother Michael and his family. At Christmas, Fred had reservations for the family to take the trip that had been postponed over the summer—a week at Disney World—but again Piper didn't want to go.

At UR, Fred began talking about the trials in his marriage with Joanne Ciulla, the ethics professor who officed next to him. Thin, with long, graying brown hair and a business-like manner, Ciulla listened as Fred explained that Piper was making odd demands, including an apartment of her own, a place to go to where she could paint. That reason didn't ring true for Ciulla, who noted that Piper had the entire household to herself while Fred was at UR and the three children were in school. "Sounds to me like she's having an affair," Ciulla said, off the cuff. Fred nodded but didn't comment. Perhaps he wasn't yet ready to consider the possibility.

As Thanksgiving approached, Piper refused to go to

northern Virginia, insisting she wanted them all to go to Texas, to spend the holiday with her family. With the credit card debt still to be paid off, Fred said they couldn't afford both a trip to Texas and one to Orlando. Piper was furious, complaining to anyone who'd listen about how ordinary a Disney World vacation was, geared for the masses, not someone with her exceptional tastes. "A bunch of us would have loved a vacation like that, but Piper acted like she'd rather have all her teeth pulled than go with Fred and the kids to Disney World," says a neighbor. That Thanksgiving, Piper went to Texas alone, and Fred took the children to visit his brother's family.

After the holiday, Piper returned from Texas sad and homesick. She missed her family and wanted to be with them. Days after returning, still suffering her malaise, Piper told Mel she'd made a decision. She was moving back to Texas.

"Your family is here, Fred and your children," Mel told her.

"I'm going to take the children with me," Piper replied. "I was so much happier in Texas with my family than I am here with Fred."

"You can't just take the children and leave," Mel advised.

Not keeping her plans a secret, Piper, Fred would later say, also announced to the children that they were moving, without discussing it with him. Yet, he'd admit they rarely talked at this point in the marriage. Frightened, perhaps sensing that events were spinning out of control, he pushed Piper to go for another round of marriage counseling, but she refused.

Even the children, who Piper claimed to love more than anything or anyone in the world, didn't appear to register with her that winter. The week after Thanksgiving, she didn't show up at Chuck E. Cheese for Callie's pizza and games birthday party. Instead Fred was there with all the little girls' mothers. When they asked where Piper was, Fred appeared embarrassed and said his wife didn't feel well.

One mother commented, "I'd have to be in the ICU to miss my kid's birthday party."

When Piper saw Dr. Welton that winter, she talked more of missing her family in Texas, saying their love had been "gratifying," and questioning again, as she had to Chris Daly years earlier, why she was with Fred. Although she'd been the one who was unfaithful, she mused that perhaps Fred was having an affair. Throughout the session, Piper was tearful and upset. "Invites some level of helplessness that is difficult to support," Welton wrote on her chart, questioning Piper's sincerity.

She must have been even more on edge than usual, because Fred would later say he called Tina that week and talked to her about the possibility that Piper was so unstable she could hurt herself. He addressed the uncomfortable possibility of having his wife committed to a psychiatric facility.

The following Friday, two days before they were to leave for Florida, Piper called Linda Purcell, a neighbor, and asked to borrow her car. Linda had been worried about Piper, who'd grown gaunt and haggard looking. In fact, Piper looked so thin and ill that some neighbors wondered if she had an eating disorder. But Linda had good memories of Piper, including how, years earlier, Piper had comforted her when their family dog had died. So when Piper asked to borrow her car, Linda agreed, assuming it was just to run a quick errand. When hours went by and Piper didn't return, Linda finally called the car phone, and Piper answered.

"Where are you?" Linda asked.

"I don't know," Piper answered, sounding strangely calm.

Later that afternoon the phone rang in Mel's house. Piper was at a doctor's office and couldn't drive. She needed a ride home. When Mel got there, Piper was nearly unconscious, barely able to walk. When Mel asked the nurse what was wrong with her, all the nurse would say was that they'd given Piper something to calm her.

At the Jablin house, Jocelyn and Paxton had just gotten home from school and were watching television when Mel and her husband, Pete, arrived with Piper. Mel was worried about the children's reactions, seeing their mother in such a strange state, so she approached them calmly and tried not to alarm them. "Your mom is okay. The doctor just gave her some medicine that made her sleepy," Mel explained. "Mr. Foster's going to help her upstairs."

As Pete carried Piper into the house, Mel watched the Jablin children to gauge their reactions, thinking they'd be full of questions, and perhaps frightened at seeing their mother so heavily medicated. But instead Mel saw no reaction at all from the children, who quickly turned their eyes back to the television. "I had the impression they'd seen Piper like this before," says Mel. "They weren't surprised."

That evening, Linda Purcell's phone rang again. She'd heard about Piper's doctor appointment and had enlisted a friend to drive her to the doctor's office to recover her car. Now Piper wanted another favor. "I need you to come to the house and tell me what you know about what happened to me today," she said. "Please, I need your help."

When Linda arrived at the Jablins', Fred, who'd rushed home after Mel notified him of the afternoon's events, looked uncomfortable. Piper, on the other hand, was manic, frenetically pacing the living room.

"Tell me what you know," she ordered Linda. "Everything."

As calmly as possible, Linda recounted what had happened, including Piper's borrowing the car and the phone call when Piper said she didn't know where she was.

With that, Piper became highly agitated. "Fred's drugging me," she shouted, not lowering her voice, although she was within his earshot. "He gave me something that knocked me out."

Purcell looked up and saw Fred staring at Piper, flushed

with embarrassment. Linda had always known Fred Jablin to be a calm, kind man, and she felt sorry for him, but she felt other eyes on them, and turned to realize they weren't alone. In the doorway, eight-year-old Paxton stood staring at his mother. Feeling sure he'd seen Piper's strange behavior and heard her bizarre accusations saddened Linda, who made an excuse and left.

From that point on Linda had an uneasiness about her neighbor and felt certain that whatever was behind the drama unfolding at 1515 Hearthglow wouldn't end well. That night, Linda recounted the events at the Jablin house for her husband. When she finished, she confided: "I'm truly frightened of Piper. She's slipping over the edge."

Early the following morning, the phone rang at Melody Foster's house.

"Where are Piper and Callie?" Fred asked. "Are they over there?"

"Fred, I haven't seen Piper since we brought her home yesterday afternoon," Mel said. "Aren't they at home in bed?"

"No. They're both gone," he said, his voice edged with panic.

. . . .

7

Piper called me in the middle of the night," says Tina. "She said Fred was drugging her. She was afraid of him, and she wanted to come to Houston and bring Callie. I told her to come."

While her family slept, Piper packed suitcases and called a limo service for a trip to the airport. By the time Fred called Melody that morning, Piper and Callie were boarding a flight to Houston. Somehow, Fred found out where they'd gone. Perhaps it was easy for him to guess. Piper had certainly made no secret of her wish to move closer to her family. If he called and asked her to return home, Piper refused. Instead, that December, Fred took Jocelyn and Paxton to Disney World, while Piper and Callie spent two weeks in Houston with the oldest of the Rountree sisters, Tina.

"Tina and Piper were exceptionally close for sisters," says Terry Wichelhaus, a tall, dark-haired former model. Wichelhaus had just moved to Houston with her then-husband, a doctor, and was in the process of building a mansion near the Medical Center when she met Tina at a party. Of all the doctors' wives there that evening, Tina was the only one to go out of her way to befriend Wichelhaus. "Tina came across as really caring, and I was grateful for that," she says. "Both Tina and Piper were that way, really showing a lot of concern for other people. They were charismatic and intelligent.

There were things about both of them that made them stand out in a crowd."

When asked about each other, both Tina and Piper gush with praise. Tina describes Piper as "brilliant, gifted, and artistic." Piper describes Tina as "inspirational, a woman who makes a tremendous difference in many people's lives, and a healer of women. She's courageous. She gets in trouble sometimes, but it's because she's not afraid to grab the status quo by the horns."

Where Piper was petite and dark, Tina, eight years older, was taller, more shapely and blond. Piper was always reed thin, while Tina had to watch her waistline. When it came to their personalities, too, they seemed opposites. Piper was flamboyant and often flighty, at gatherings bubbly and fun, while Tina was intense. Like Piper, Tina had a presence, but hers was a serious demeanor, a take-charge, motherly attitude. The connection between the two sisters was strong. Those who knew them say that Tina had a power over her younger sister, that when Tina spoke, Piper listened.

In Houston, Tina, a nurse practitioner, owned and ran the Village Women's Clinic in an eclectic neighborhood not far from Rice University. With a master's degree in women's medicine, and under the supervision of a physician, Texas law allowed Tina, who'd begun her career with a nursing degree from UT, to oversee patient care much as a general practitioner would, giving examinations, doing procedures, and prescribing medications.

In an old converted house on Bissonnet, a busy street in a neighborhood of older homes in what's often referred to as Houston's Rice Village, Tina and her staff dispensed medical care and counseled patients. By her own admission, the business side of the clinic, filling out forms and paying bills, held little interest for her. It was patient care that she loved. In turn, many were devoted to her. "She's this really amazing woman, who you just know cares about you as a person," says one patient, who has been with her for many years. "Tina has a heart of gold."

The Village Women's Clinic didn't accept insurance, and many of the patients who went there didn't have any. Out front a large sign with a picture of a smiling Tina in a white medical jacket with a stethoscope around her neck listed her specialties: well-woman and school exams, weight counseling, emergency contraception for the morning-after pills she prescribed, and natural menopause counseling. Inside, the clinic was warm and inviting, decorated in antiques and Asian rugs.

Tina lived just blocks away, in an old gray Victorian cottage, on a corner lot on Albans Street not far from the acres of hospitals that comprise the Texas Medical Center and Houston's downtown business district. In the past decade, homes in the neighborhood had been sold and demolished, leveled to yield lots on which the upwardly mobile built imposing two-story homes. So much square footage on such small lots often left only borders of trees and small patches of grass. In contrast, Tina's yard was large and overgrown with vegetation. Wicker furniture waited on an inviting front porch, and in the living room she had a grand piano. Through the years, Tina gained a reputation for being an eccentric neighbor. When she complained that tree trimming companies charged too much, she climbed the thick, dark branches of her yard's live oak trees with a saw. When Tina didn't like streetlights the city installed, legend had it that she took a gun and shot the bulbs out. "When she digs in her heels she really digs in her heels," says a man she dated. "You can't tell what will set her off or why, but when she gets stubborn, Tina can be famously stubborn."

At Christmas she gave parties that included horse-drawn carriage rides through the streets of Houston. Up to one hundred people packed her house, as her German shepherd meandered through the crowd. All the while, Tina drank amaretto and circulated, pinning down guests for intimate conversations. She had a strong intellectual curiosity and often psychoanalyzed those around her. She was habitually late, even for patient appointments at the clinic, but when she arrived, she entered a room like a whirlwind. And those who knew her would say that like Piper, Tina Rountree knew how to live life to the fullest, to suck the pleasure out of every moment.

Later, Tina would say that when Piper arrived in December 2000, she and Callie were both ill and exhausted.

"The first seventy-two hours, Piper slept," Tina says.

She nursed them, she contends, weaning them back to health. Piper told her about what had happened at the doctor's office the day before, describing the injection she'd received as Valium to treat a migraine. She said Fred had been furious, that he'd ripped her clothes off and put her to bed, calling her an embarrassment. Fred, she charged, had been behind the entire incident, claiming, as she had the night before to Linda Purcell, that he'd drugged her.

A troubled marriage was a subject it could certainly be expected that Tina would identify with. Her first marriage to an attorney yielded two sons and ended in divorce. Only two months later she married Dr. Praver, the Houston gynecologist she'd later claim had dispensed Prozac to Fred Jablin. Praver's marriage of thirty-four years to his wife, Maureen, had ended abruptly about the same time Tina's had. "We had an Ozzie-and-Harriet-style family," says Praver's daughter, Janet. "All of a sudden, Tina was in the picture and our lives as we knew them were over."

On July 1, 1990, one week after Dr. Praver walked out on Maureen, she committed suicide. Their oldest son, Rick, was so incensed he sued his father. During the height of the family chaos, in 1992, Tina wrote a letter to Rick. In it, she psychoanalyzed the Praver children, chastising them for not accepting her in their lives. "It's like you're all stuck at three years old," she wrote. "Someone takes your toy away and you either hit the kid back or just withdraw from the big mean thing that took your toy away." She never acknowledged that what they had lost wasn't a mere toy, but their mother. She closed by writing: "At least have the guts to tell me to my face you think I'm a blond bimbette after your dad's $."

The lawsuits were dropped after Rick, too, committed suicide.

Howard Praver married Tina Rountree Gano in August 1990. Friends say that he called her "Sunshine" and had their towels embroidered with yellow suns. Yet the marriage ended after five years. Their parting was bitter. She claimed in court papers that he'd been physically and psychologically abusive. "Never happened," he'd say years later. Again playing psychiatrist, she wrote to the court, "I was concerned for his mental condition."

Three years after the marriage to Praver ended in divorce, Tina Rountree Gano Praver was dating Grant Heatzig, a geologist who ran a small oil-related company. A tall, good-looking English transplant, Heatzig had been diagnosed with leukemia years earlier but appeared to be in remission. He and Tina dated for months and then broke up. Friends say that when he suffered a stroke and was diagnosed with a brain tumor, she came back into his life.

In October 1999, eight days after brain surgery, Heatzig would later contend, Tina invited him to Las Vegas. When they arrived, he learned that she'd asked her family and friends to meet them there, to witness their wedding at the posh Bellagio Hotel and Casino. He'd claim he knew nothing of the plan, but that he cared about Tina and decided to go through with the ceremony. Terry Wichelhaus had only two hours notice to get on the airplane for Las Vegas to be at the wedding. "Grant was brilliant, a supreme human being, but he was so ill," she remembers. "We all knew he was dying. We thought maybe the marriage would be good for him. Tina was a nurse and she could take care of him."

It didn't turn out that way. Tina hired caregivers to watch over her husband, day and often into the night. He grew lonely and angry, and six months after they married, in March 2000, filed for an annulment. Tina countered and filed for a divorce. In his annulment application Heatzig wrote: "It is my opinion that Tina induced me into the civil marriage ceremony in Las Vegas solely for the purpose of attaining financial gain for herself without any real intent of acting as my wife."

If Tina had hoped to cash in on Heatzig's wealth, in excess of a million dollars, that didn't happen. In the end she received little, just $50,000 and his membership at the Houstonian, the stylish hotel and athletic club in Houston's posh Galleria shopping district, whose members are success-driven young professionals and Houston's movers and shakers. It was a Houstonian hotel room that the first President George Bush listed as his official Texas address during his years in the White House.

The way the marriage ended, Terry and others sided with Heatzig, and their view of Tina changed. Instead of a caring woman, they thought they saw another side of her, that of a calculating woman willing to take advantage of a dying man. "Grant was bitter about Tina," says Wichelhaus. "It was so cold, making his last year so troubled."

Six months after the marriage officially ended, in December 2000, Heatzig died. That same month, Piper arrived on Tina's doorstep. "I wasn't just her sister, I was Piper's best friend," says Tina flatly. "Of course, I took her in."

Perhaps it was as much for the children as her sister. "Tina didn't have any girls, so Callie and Jocelyn were extra special to her," says Glenda King, a close friend. "Tina thought of herself as more than an aunt."

The way Tina would later tell it, Callie ran a low-grade

fever during the beginning of the visit. She'd diagnose it as a minor influenza. And Piper was suffering from exhaustion and more. "Her spirit had been beaten down," says Tina. "She was so dependent on Fred, by his design, she couldn't do anything. I worked with her, told her she was strong and independent, that she needed to get out of the situation. Before long the old Piper returned, the spunky Piper. Fred was undermining her self-esteem."

After his week at Disney World with Jocelyn and Paxton, Fred flew to Piper in Texas—as he had seventeen years earlier when he flew to Germany to convince her to return to him. He brought both the older children with him, in his bid to reclaim his youngest daughter and his wife and reunite his family.

Days later he and Piper returned to Virginia, Fred would say later, with the agreement that he would find a teaching position that would bring them back to Texas. That winter he did, in fact, apply for a position at UT, in the communications department where he'd once worked with Daly. Perhaps Piper intended to do the same, to make changes and to try to keep their family together. Later the evidence would suggest otherwise, and that what was on her mind that winter was money.

Months earlier Fred had taken nearly all Piper's credit cards away, but that December she maneuvered to get them back. When Fred's Purdue MasterCard disappeared, he reported it lost. Later he'd charge that Piper then intercepted the replacement card in the mail, along with a new CitiBank card she ordered on his account. In less than thirty days Piper would charge more than $9,000 in cash advances and purchases of $10,454.77 on Fred's credit cards, along with transferring $3,000 from his paycheck into an individual account she opened at a Wachovia bank. What Piper bought on the cards would later make it seem that she was building a nest egg for her future, including clothes, tennis equipment, three years' worth of prepaid hair salon services, and $7,000 in prepaid moving services, more than enough to move her home to Texas.

It would be nearly a month before Fred would learn of the new debt, but the strained marriage still barely lasted through the holidays. To Fred's great sadness, just after the New Year, on January 6, 2001, Piper packed her bags and moved out of the Hearthglow house. By then her friend Loni was in the midst of a divorce. Piper took advantage of having a refuge to flee to and moved in with Loni.

For the next few months it seemed Piper wasn't sure what she would do, as she moved back and forth between Loni's home and the house on Hearthglow Lane. Loni would recall that period as great fun, saying that she worked while Piper stayed home and kept the house and cooked dinners. During the week, Fred had the children, but on weekends they joined Piper at Loni's. On her days off, she and Piper took the children to Richmond's Maymont, a parklike setting on the James River, with Japanese and Italian gardens, a Victorian mansion, and a petting zoo. There, they picnicked, the children played, and Piper painted. In the evenings the women drank wine and talked. "We did bubbles, putting the kids in the tub together," says Loni. "It was fun, and everyone got along. It was a wonderful time."

When Piper circulated back home with Fred, the scene wasn't as tranquil. Instead, on January 11, Piper called the police, complaining that during an argument Fred pushed her into a wall. She did, in fact, have a small cut on one hand. Yet when the officer asked if it was from the argument, Piper said she didn't think so.

Later, Fred, who'd grown so thin he appeared only slightly larger than his diminutive wife, would tell others that Piper had pushed him, and that it had shocked him that she could be physically aggressive. But that day, to the police, he described the argument as purely verbal. Looking for the truth, the police questioned the Jablin children, who'd been there throughout the confrontation. They agreed with their father, that there'd been no physical violence. On their forms, the police noted that there were no signs of physical abuse.

Still, Piper was adamant that Fred was abusing her, and she told police that Melody Foster "knew what was going on and could verify it."

Before leaving the Kingsley subdivision, the police took the additional step of driving around the block and knocking on Melody's door. When they told her why they were there and asked her if she'd ever seen any sign that Fred Jablin was physically abusive with his wife, she answered, "Absolutely not."

"My impression was that the police thought the same thing I did, that Piper was unstable and making it all up," Melody would say years later. "They didn't believe Fred had done anything to her."

Increasingly, Melody worried about her back-door neighbor, seeing Piper as wound so tight she could easily snap.

That spring, when Mel asked Piper to help her line a chest with fabric, Piper agreed. But that day at Mel's house, Piper was anxious, jittery, and tense. As Mel watched her warily, Piper popped two Adderall stimulant pills in her mouth. Although she had her own prescription for the pills from Dr. Welton, it appeared those weren't enough.

"These are Joce's," Piper said. "Have you ever tried Adderall? They're great."

Having witnessed such chaos in the house, their parents' marriage imploding before them, the Jablin children must have wondered what would happen next. On January 17, six days after the police were at their door, Jocelyn, by then an eleven-year-old with dark blond hair and glasses, told her therapist that the thing that would make her the saddest was if her parents divorced and lived apart. What would make her the happiest, she said, was to have her family remain intact.

"But that's probably not going to happen," she admitted.

The next day, Fred wrote Piper an e-mail at 8:43 in the morning. The subject was marked as "Everything." It was a touching letter, in which he begged to keep their family together. "I realize that you and I are about as angry with each other as possible and that we are both hurting badly. First, despite all that's happened, I still care very much about you . . . a candle of love still remains burning, although it appears on a trajectory to burn out. Second: I miss you. Third: It is clear we have both failed one another in many respects . . . I guess I am looking for a miracle . . . You seem to believe in them and have noted that your life has never followed the norm . . . what my heart and soul seem to be telling me (or perhaps what you might consider my angel communicating with me) . . . I feel convinced that we are not working our problems out in the best manner."

As the letter unfolded, Fred suggested they remain married yet stay in their separate bedrooms while they consult a marriage counselor. In the meantime, he promised to continue to look for a position in Texas. If he didn't find one, he offered to take a leave of absence from UR in July and move to Texas with Piper and the children, so that she could be closer to her family. "All of this may take some kind of miracle, but miracles do happen, especially if people want them to happen. My heart and soul keep reminding me not to forget this, and I hope yours do as well . . . Love, Fred."

Whatever his intentions and Piper's reaction, the relationship continued to falter. Two weeks later, on the evening of January 31, the police were called to Hearthglow Lane a second time. This time Fred met them at the door. "My wife's acting strange, and she's trying to take my kids away," he said.

Once inside, the police found Piper agitated, complaining that Fred had the children in one of the bedrooms with the door locked and wouldn't let them out. She'd told him she was leaving and wanted to take them with her, but he refused to let the children go. By the time the police left, they'd noted in their report pad that there were no signs of any physical aggression between the couple, and that they'd both promised to stay apart for the rest of the evening and talk the matter over in the morning.

On February 4, Fred and Piper's feud smoldered. By then he'd discovered that she had intercepted his credit cards in the mail and driven up the family debt by an additional $20,000. At UR, Fred told Joanne Ciulla about Piper's reckless spending. "He couldn't understand what she was doing," says Ciulla. "It was like he was still trying to make sense of it all. He kept asking why a mother would do such things." Furious, Fred threatened to press charges against Piper for mail fraud.

The next day the madness continued. Piper took Jocelyn to an appointment with her surgeon, Dr. Gable, and then checked into a hotel with the children, calling Melody and telling her that she was afraid Fred would have her arrested as he'd threatened. Mel explained that it was unlikely that even if Fred pursued charges the police would become involved. Frantic, Piper appeared not to hear her.

Two days later Piper was still in hiding at the hotel, telling Melody that she expected the police to pick her up at any moment. When Piper called her psychiatrist, Welton, to cancel her appointment, she told him that she was afraid Fred might have police waiting to arrest her at his office. On her chart, Welton noted that Piper's actions were "part of a pattern of chaos." At this point in the therapy, he noted that Piper's emotions were out of control, and he questioned her ability to use medication to control her mood swings.

That day, while Piper didn't show up at Dr. Welton's, she

did walk into the Henrico County magistrate's office to swear out a restraining order and an arrest warrant against her husband. On it she charged that on the night of the initial police call to the house, nearly a month earlier, Fred had been physically abusive. If Fred thought that he and Piper couldn't have hurt each other any more than they already had, he didn't anticipate what happened next.

Hours later police arrived on the University of Richmond campus and walked up to the imposing building that housed the Jepson Leadership School. Inside the heavy oak doors, they entered the hallowed corridors where such lofty concepts as morals and ethics were taught, with a warrant in their hands for the arrest of one of the institution's most honored faculty members, a world-renowned expert in his field, Professor Fredric M. Jablin. The charge was domestic violence. As Fred was confronted, arrested, and walked under police escort from the building, he hung his head, and his colleagues and students looked on in amazement.

Later that same afternoon, Melody's phone rang. It was Fred. He sounded tired and embarrassed, and at the same time mystified, as if the unfolding drama couldn't possibly be happening to him. With a restraining order in effect, he wasn't allowed anywhere near Hearthglow Lane. Despite all he'd been through that day, including being booked at the police station, Fred remained calm. He asked Mel, who had a key, if she would retrieve his clothing and his allergy medicine from the house and bring him a suitcase. "Fred was stunned," Mel would recall years later. "He said, 'It's like my life is turning into a bad Lifetime movie.'"

That afternoon, Fred moved into a Studio Plus motel a few miles from the university, and with Fred barred from Hearthglow, Piper and the children returned to the house. Later it would seem that the die had been cast, and that the circumstances that would eventually lead to tragedy were falling in place. There was no denying that Piper Rountree and Fred Jablin were becoming increasingly involved in what would become a destructive game of strategy, control, and revenge, and their children would soon become the battleground.

"It was beginning to look like that movie, *The War of the Roses*," says a friend. "They were both out for blood."

On March 11, Fred, accompanied by an attorney, went before a juvenile court judge asking for an emergency hearing. Fred argued that the children were unsafe with Piper and asked for temporary custody. The judge, after listening to the evidence, agreed.

When Piper found out, she was fuming. The next day she retaliated by returning to the magistrate's office to swear out a second warrant for his arrest, again on a charge of domestic violence. This time, while Fred was maneuvering through the system, trying to be released, Piper used some of the $7,000 in moving expenses she'd prepaid on Fred's credit card and took what she wanted from the house, including the master bedroom and dining room furniture, an antique pottery set, and Fred's mother's piano, pearls, and wedding ring. She then moved into a town house on Castile Place, a short drive away, but didn't tell Fred.

Despite the growing cloud of anger forming around them, Fred was not ready to give up on the marriage. He called, begging Piper to take him back and reunite the family, if not for their sakes, for the sake of the children. "Even once Fred understood the scope of it, he didn't know what to do," says Melody. "He loved Piper. As crazy as it was, he loved her."

Piper refused.

Perhaps reluctantly, on March 16, Fred instructed his attorney to file for divorce on the grounds of desertion. In the paperwork, he said he saw no possibility of reconciliation. That day, with Piper gone and living in her town house, the judge lifted the restraining order, allowing Fred to move back into the house on Hearthglow Lane and again ruling that Fred had temporary custody of all three children. From that day on the battle lines were clearly drawn, directly through the hearts and minds of Jocelyn, Paxton, and Callie.

On one hand, the three children had a father intent on making sure they were safe. And after years of watching Piper grapple with motherhood, Fred had come to the conclusion that his wife could be dangerous. "Fred saw Piper as irrational and too distracted to be responsible for the children," says Daly. "He saw the credit card bills as a violation of trust, and feared she'd run off with the children and simply disappear."

At the same time, Piper envisioned herself as an exceptional parent, the very essence of motherhood. She would never, could never, give the children up willingly, for they constituted everything she believed about herself. "My children always came first to me," says Piper, with such emotion that her hands begin to shake. "If I'd wanted to, I was incapable of giving them up. Why would I? It was in their best interests to be with me. I was their world and they were mine."

Ironically, just as Fred's private life dissolved around him, his professional life again surged forward. That month, the dean of the Jepson School was forced out, and Fred was tapped as acting dean. The job came with a $30,000 raise, one sorely needed with the court's order that he pay Piper $1,167 a month in temporary support. At home, Fred hired Ana, a petite, dark-haired woman from Brazil, to be the children's nanny. She was a gentle person, and they quickly grew to love her as a part of the family, but that wouldn't mean that any of their lives would take on any semblance of normalcy.

There was no denying that Piper was angry and hurt, and the rest of the Rountree family, especially Tina, was furious. "It's all they talked about," says Terry. "The Rountrees couldn't believe that Piper didn't have her children. To them, it was simply impossible."

That spring, 2001, Piper filed a cross-complaint, asking for her own divorce, and charging Fred with cruelty, both verbal and physical abuse. She asked the judge for joint custody. "I was born to be a mother," she'd say later. "It was in my blood. He had no right to take my children from me. Absolutely no right."

Fred didn't agree. As he'd habitually read and familiarized himself with the work in his field, he now studied Virginia's divorce laws, determined to make sure his attorney did everything possible to ensure that he retained custody. Toward that end, in April, Fred's attorney subpoenaed Piper's medical records, including those of her sessions with her therapist, Dr. Welton. In response, Piper's attorney filed a motion to quash Fred's subpoena, to prevent such potentially damaging material from being used against her.

The bad news kept coming. As the school year drew to a close, Piper took the children out of school for a trip to the Blue Ridge Mountains, hiking with Tina. What galled Fred was that Piper did it on the day Jocelyn and Paxton were scheduled to take the Standards of Learning tests, Virginia's statewide standardized assessment tests that they had to pass to be promoted to the next grade. "Why would she do that? She's their mother. Shouldn't she know better?" he asked.

As much as he'd loved Piper, as dedicated as he'd been to her, her erratic behavior was chipping away at the last remnants of his attachment to her.

If Fred had any doubt that the marriage was over, that ended on June 8. For months he'd speculated that Piper was involved in an affair. That day he had his suspicions confirmed. "Fred discovered that Piper was having an affair with Jocelyn's doctor," says John Daly. "It's true that Fred had his rules, but normally he forgave people who broke them. He rarely got angry. Instead, he'd try to help people, convince them that there was a better way to do things. But adultery, that was the biggest rule of all, and Piper had broken it. Not only that, but she took their children to spend time with this man. That, Fred couldn't forgive. He was as dedicated to the children as he'd ever been to Piper, and he was determined to protect them from her. That's when the real war began."

. . . . .

8

Fred wanted to go after Piper's lover," says Melody. "He threatened to file complaints with the medical board and try to take away his medical license. As he saw it, the doctor had violated a trust, and he wanted revenge."

The revelations kept coming throughout the summer of 2001, when Fred learned, in February, that Piper had been with Gable at Kingsmill Resort, a classy spa just outside Colonial Williamsburg, and in Toronto with him at the end of April. Even more damning in Fred's eyes, Piper had taken all three children on their dates, hiking, roller skating, and out to dinner. As angry as Fred was—and he was livid—he also appeared mystified by what seemed a bizarre turn of events: The doctor was thin with an angular face, a slightly built, balding Jewish man.

"The strange thing is that this guy looks just like me. Piper left me for a man who looks enough like me to be my brother. Why would she do that?" he asked.

Mel had no answers.

At UR, Fred cornered his colleagues in the hallways and their offices, again detailing the odd turn of events in his life. When he pigeonholed one of them, he'd talk for upward of an hour, dissecting the dissolution of his marriage, trying out ideas on them, and asking for advice. "I've seen this doctor and he looks a lot like me," he told Joanne Ciulla one afternoon. "Can you believe that?"

Amazed, Ciulla began laughing. Despite all he'd been through, Fred smiled, then caught up in the absurdity of it all, laughed with her.

Yet, Fred Jablin never truly saw the situation as funny. Instead, he described the affair as even more evidence of his wife's instability. Piper was acting without bounds, he said, not thinking of their children's needs, and that was something he couldn't tolerate. "Fred saw the affair as even more reason why he needed to have custody of the children," says Mel. "It made him determined to win in court."

At home, he shored up the fortress he was building around his children, to protect them from their own mother. Throughout the eighteen years of his marriage to Piper, Fred had rarely attended synagogue, and they'd celebrated both Christian and Jewish holidays. With her out of his life, he returned to the religion of his childhood, seeking the strong foundation his faith offered. That summer he signed the children up for religious classes at Congregation Beth Ahabah, in downtown Richmond. In an impressive stone and brick structure with classic columns, the synagogue, founded in 1789, was the oldest in Virginia and the sixth in the nation. Outside, a plaque read: "What doth the Lord require of thee: Justice—Mercy—Humility. Micah 6:8."

To get through this heartbreak in their lives, Fred relied on religion, friends, and family. His brother Michael lived just outside Washington, D.C., with his wife Elizabeth and their two children. After the split, Fred and Michael talked each Sunday on the phone, and on holidays Fred traveled to visit him with the children. That year Fred wrote a new will, and Michael became Jocelyn, Paxton, and Callie's guardian in the event of Fred's death. If something happened to him, Fred didn't want Piper to have control of the children or their money. And he did worry about something happening to him: As a precaution, he had a security system installed on the house. In what had to have been a sad task, he talked with the children about their mother, explaining that Piper was unstable and that he feared she would take them and disappear. After many such conversations, he told friends he'd grown to believe that Jocelyn and Paxton understood and would watch out for themselves, but he worried about Callie. "Fred saw her as vulnerable," says Mel. "He understood how troubled Piper was, and he feared she could become dangerous. He didn't know what she might do."

Fred's anxiety grew when he heard from Dr. Gable's wife, Elaine. She, too, had learned of the affair, and she suspected Piper was behind something odd that had happened. One day when Elaine went to her mailbox, she opened it and found what appeared to be a snake. Her heart pounding, she slammed it shut. Only later did she learn that the snake was a rubber toy. Elaine didn't interpret the scare as an innocent prank but as a threat, and she told Fred that she felt certain Piper was responsible.

Meanwhile, Piper acted as if she hadn't a care in the world. She dated Gable, introducing him to friends, and telling many that she thought he was "the one." That the physician was wealthy with a thriving practice probably didn't diminish his appeal.

Rather than looking for work, Piper concentrated on her art.

That summer, like cities across America, Richmond sponsored an art exhibit by local artists. Chicago had cows, Houston long-horned steers, and Richmond chose a large, preformed resin and polyurethane Atlantic striped bass, and its exhibit was called "Go Fish!" Artists throughout the community volunteered, and Piper was among them. She spent much of that summer in a friend's garage, working on the project. When the kids came for visitation, they helped. In the end, *Soul Rays,* as she named her bass, was painted in blocks of vivid color, with a black fin. Sponsored by Noah's Children, a children's hospice, Piper's fish took its inspiration from children in grief counseling. Inside glass blocks, inserted like scales, she preserved the children's memories and thoughts, in their own words. A red rose inside one block was dedicated to a dead mother who'd loved flowers. Sand preserved a young girl's memories of a deceased sister who'd taken her to the beach.

One of *Soul Rays*'s glass scales would later haunt those who noticed it. The boy it was dedicated to was identified only as Ryan, age sixteen. Scrawled across the glass in black marker Piper had written: "I miss my dad because he shot himself." Inside that block was a small grave with a tombstone marked R.I.P., and suspended in the glass was an empty brass cartridge, from a fired bullet.

Fred and the children were at the beach house in North Carolina on July 1, when Piper arrived at the Hearthglow house at 2:43 in the morning. Later she'd claim she'd come to visit the family pets, her much-loved cats, ferrets, and dogs. The new security system horn bellowed when she opened the door, and police responded. When Fred was contacted by the officers, he told them to arrest Piper for trespassing. They refused, saying the house was in both their names and that there was no court order barring her from entering it. When he returned home, he'd later say, more of their possessions were missing, including the Kiddush Cup he'd inherited from his parents, a Jewish ceremonial cup used in Sabbath celebrations.

That afternoon, twelve hours after leaving Hearthglow Lane, Piper sent an e-mail to a friend in which she mused about her relationship with Gable, saying that he was trying to "reconnect" with his wife. She joked about the alarm incident of the night before, laughing that she set the alarm off two more times when she let the cats out, gleeful that Fred could be billed for the police responding.

"I'm trying to stay inside to avoid falling comets," she wrote.

That night, the friend responded, writing: "A man can not have two real bitches in his life at the same time." It seemed to predict that the doctor could not tolerate his wife and Piper simultaneously.

With the battle over custody intensifying, Piper hired a new attorney, and so did Fred. He chose Susanne Shilling, a well-known Richmond divorce lawyer who officed in a converted house not far from the historic St. John's church. Shilling, known as an expert in custody cases, sat down to talk with him and heard his story. Fred would later tell a friend that he instructed Shilling to proceed as quickly and as aggressively as possible, to ensure that he win custody of the children. The day after Piper's late night visit to Hearthglow, Shilling began her assault, redrawing Fred's divorce petition, changing the grounds from desertion to adultery.

The stress must have been incredible that spring and summer. Most neighbors would say they never saw Fred Jablin lose his temper, but one day Annie Williams, Jocelyn's soccer coach, gave her a ride home when no one showed up to pick her up from practice. When the Hearthglow house was dark and deserted, Williams wanted to leave a note on the door to tell Fred where Jocelyn was and take her home with her. The twelve-year-old became fearful, crying and insisting she was supposed to go inside and wait. Just then Fred drove up. He was livid, shouting at Williams, "What are you doing with my child?"

Although Williams assured him that everything was fine, that she was just giving Jocelyn a ride home, Fred continued to fume, "No, it's not fine."

Stunned at his reaction, Williams pointed out that she had done the responsible thing when no one had come for Jocelyn after soccer practice. Williams then offered to give Jocelyn rides home from future practices, believing it would help Fred, who'd been struggling to balance work with the children's needs. He turned her down, defensively saying, "Never mind. I've got it covered."

Williams had seen Piper looking glazed, like she was on heavy medication, and Jocelyn's soccer coach left the Jablin house that day worried about the children, concerned that neither of their parents were in the frame of mind to raise them.

Yet, others in the neighborhood experienced a very different Fred that year. They saw him biking and throwing the ball around in the front yard, laughing and playing with the children. Fred had always been single-minded. Now that Piper was gone, he focused his attention on the children, as he'd once focused on her.

In the court system, the Jablin divorce was assigned to Judge Catherine Hammond, a middle-aged woman with short, windblown blond-brown hair. She appointed a guardian *ad litem* for the three children, a woman named Donna Berkeley. Fashioned under old English law, in Virginia divorces were handled in two parts. The first, overseen by Berkeley and Hammond, would consider the children's custody. The second part of the process would distribute assets, and for that Hammond appointed Edwin Bischoff, a commissioner of the court.

There was a lot to lose for both Piper and Fred: not only custody of their children, but their money and property as well. Yet, in hindsight, it seemed that Piper wasn't concerned. In fact, she appeared indifferent to the pending divorce case, as if her actions were immune to scrutiny.

Throughout the summer, Fred received e-mails forwarded to him through others and coming directly from Piper. It was odd. She was an attorney, a bright woman, yet she wrote to friends and copied everyone on her list. Sometimes the material was potentially damaging to her bid to reclaim the children, like on June 8, when an old friend from Texas e-mailed, asking Piper to ask her angels where her lost dog could be found. Piper replied with obtuse advice: "Seeking is what is not necessary. Listening is the key."

At times she was flippant, like the day she e-mailed nearly everyone in her address book—including Fred and Paxton—with an idea for a new type of computer joystick, suggesting it could be used for sexual gratification. At other times she was mournful, lamenting the new sadness in her life, as on June 13, when she e-mailed her family and friends. The subject: "Night Moans." In the body of the e-mail she wrote: "Crying myself to sleep missing my babies." She complained that Fred had dragged a sick Callie to Jocelyn's piano concert. She said when she arrived, Callie wanted to sit on her lap. "I lean so heavily on Paxton for strength and love . . . and he me," she wrote.

At the same time, e-mails flew back and forth between Fred and Piper, in which she chastised him for interfering with her visitation when he called the children at her apartment. Their relationship worsened, and the plans to hand off the children to one another for visitation became increasingly complicated. Once, Piper picked up the children from swimming lessons to take them to an afternoon with her paramour. Fred was furious.

On July 18, as the tension mounted, Fred filed a petition with the court, this time asking not for joint custody, as he and Piper had discussed in the past, but for full custody. His reasons: Piper's erratic behavior, including her bouts with depression, her inappropriateness with the children, and her contention that she spoke to angels. To back him up, he produced copies of her e-mails. Fred also claimed in the petition that Piper had falsely charged him with domestic violence. By then the charges against him had been dismissed when Piper failed to appear at a hearing. Fred contended that the entire episode had been a ruse. He charged that she had sworn out the warrant simply so she could "loot" the Hearthglow house while he was in jail.

"The children are suffering, confused by their mother's behavior," Shilling argued for Fred in the petition.

The night the petition was filed, police again responded to 1515 Hearthglow Lane, while Fred and the children were at the beach house. This time the locks had been changed, and Piper broke a garage and a kitchen window to get inside. Again police called Fred and advised him that there were no orders on file to prevent her from entering the house. When Fred talked to one of his old graduate students, he described Piper, a woman he'd once dearly loved, a woman who'd once loved him, as evil.

"She's trying to take advantage of everything she can," he complained. "She's totally unscrupulous."

At times Fred walked through the backyard gate over to the Fosters, as Piper had once regularly done, to watch the children play and talk to Mel. She listened sympathetically as he lamented the turn of events that had changed his life. He worried that the divorce judge wouldn't clearly see Piper. In the courtroom, he complained, she appeared professional, a lawyer and mother concerned about her children, but while she had the children, she seemed oblivious to their needs and their safety. Like the time she allowed Callie and one of the older children to cross a busy four-lane road near the apartment, to go to a convenience store. Word spread of Piper's reckless behavior, and some of Paxton's friend's mothers were so concerned about their children being with Piper that before they'd allow them to stay overnight with Paxton, they called Fred to make sure the children wouldn't be going to Piper's apartment.

During one of her talks with Fred, Mel admitted that she and other neighbors had seen signs that Piper was troubled going way back, and she told Fred of the rumors that Piper had affairs with some of the young men she played tennis with at the club.

"Why didn't you tell me before?" Fred asked.

"We all thought you knew or didn't want to know," she said. "If we'd told you, would you have believed us?"

Fred thought for a minute and admitted, "Probably not."

The first courtroom custody battle was scheduled for August, and both Piper and Fred enlisted family and neighbors to be witnesses. Piper asked Annie Williams, who refused. She couldn't forget the glazed look she'd seen on Piper's face, and she worried that perhaps her friend was taking drugs. By then Piper was threatening to also charge adultery in the divorce, although she offered no evidence that Fred had been unfaithful. In her e-mails to friends, she fumed, angry at the turn her life had taken. It seemed apparent that she hadn't envisioned the divorce turning out as it had thus far, that she'd have little money and that Fred would be pushing for custody of the children.

"My life is a mess," she wrote. "Generally, I am screwed."

Although Piper had written two résumés, one touting herself as an artist and the other as an attorney, she had yet to find a job. Fred told many that she wasn't truly looking, and a sentence in one of her e-mails suggested he was right. "If I take a job during the summer while the kids are out of school it will be held against me," she wrote, regarding custody. She'd offered to give art lessons to the neighborhood children in Kingsley, many her own children's friends, but at $350 a week per child. "It was insane," says one neighbor. "Why would she think any of us would be willing to do that?"

On the flyer she distributed to drum up students for her art classes, Piper wrote: "It's about nurturing. It's about needs. It's about soul and spirit. It's about color, light, and love." Meanwhile, on the Internet, she put her artwork up for sale on a Yahoo website she named Light Worker Galleries.

When someone asked, she defined a light worker as a type of mystic, someone who acted as an intermediary between the physical and the spiritual worlds.

Each time Jocelyn, Paxton, and Callie went with their mother, Fred stewed, worried about what could happen. When she took the children out in a boat without life preservers, he called it more proof of her bad judgment. "Why would a mother do that? Why?" he asked everyone who'd listen. Incensed, he redoubled his efforts to gain sole legal custody, filing more motions against her, using more of her own e-mails as evidence, including the tawdry comment about the joystick.

In Austin, when Margaret Surratt, who worked in the UT office, heard about the divorce, she felt sorry for Piper, remembering how single-minded Fred could be. She had no doubt that he would be a tenacious opponent in a divorce. "Putting those two personalities together over time had to be volatile," she says. "Piper was a butterfly, and Fred was a bloodhound."

In Harlingen, meanwhile, rumors circulated about Piper's divorce, especially whispers questioning why a mother would lose custody of her children. Old high school friends heard the gossip: that Piper had a physician lover who'd offered her the world, and all that stood in the way was Fred, who refused to let her have the children. "The entire Rountree family was up in arms about it," says a friend. "They saw it as unfathomable that Piper, who was this wonderful mother, would lose custody."

The custody hearing on August 6 was contentious throughout. Tina and Piper posed poster-sized photos of the children with Piper and other members of their family in the gallery seats, a way of displaying for the judge the happy family the children enjoyed with the Rountrees. On the stand, two of Piper's friends testified that she was a good mother, the kind who cheered for her children on the sidelines during soccer games. But others testified otherwise. Melody had struggled with whether she'd take the stand on Fred's behalf. She thought back to her own experiences as a mother and felt a sense of compassion for Piper, whom she'd once considered a friend. Despite everything, she knew that Piper, in her own way, loved the children more than life itself. All mothers make mistakes. Was she being too hard on Piper? she wondered.

And there was something else: She asked Fred if he thought testifying could put her and her family in danger. She'd grown to believe Piper was ill, and she didn't know what she might do.

"I can't tell you that you'll be safe from Piper. I don't know what she'll do anymore," he answered, his brow knotted in worry. "I fear that someday she might do something that unintentionally hurts the children. Every time they go with her, my heart is in my throat."

In his darkest thoughts, he admitted, he feared that in one of her depressions she might kill herself and, not willing to leave them behind, the children.

In the end Melody decided that despite the risks, helping Fred keep custody was the right thing to do.

In the courtroom, before the hearing, Piper approached her saying, "I'm glad you're here, Mel. Now you can help straighten this mess up," as if she didn't realize Melody was Fred's witness, not hers.

On the stand, Melody told the judge everything she'd observed, including Piper's inattention to the children and her erratic behavior. She could feel Piper's eyes on her. "I kept my eyes straight ahead," she says. "I didn't look at her. The children worshipped Piper, and she was a lot of things to those children, but she wasn't a good mother."

When it was his turn on the witness stand, Fred told of the uncertainty in his life since the separation. That same night he had another example, when it appeared that Piper had again entered the Hearthglow house. After taking the children out for dinner and a movie, Fred returned home to find his mail not in the mailbox, but inside the house on a table. He made a note of the event on his computer, adding it to a long list of Piper's transgressions: "Entered house and went through mail." Each day, he meticulously recorded her sins and omissions: the birthday cake she promised to but never baked for one of the children, the days when she had custody and refused to take them to their therapy sessions, dance lessons, and soccer practices.

The following day brought more bedlam, when Tina showed up at the Hearthglow house unannounced and stormed into the front yard. The nanny, Ana, was outside with Paxton and Callie, and when Ana confronted her, Tina pushed past her and went upstairs to see Jocelyn. Ana called Fred, who talked to Tina on the telephone, ordering her to leave. Tina hung up on him. He rushed home, but by the time he arrived, Tina was gone, leaving the children in disarray. Jocelyn was sobbing, Callie appeared confused, and Paxton, upset, spent the rest of the evening in his room.

At the University of Richmond, Pulitzer Prize–winning biographer James MacGregor Burns joined the Jepson School as a senior fellow, and Fred, as acting dean, welcomed him aboard. Meanwhile, Piper fired her second divorce attorney, one of the finest in Richmond, and wrote terse letters to Fred's attorney, Susanne Shilling, claiming she was holding improper conversations with the judge and offering her instructions on how to be ethical. When Fred dropped the children off that weekend, Piper again called the police, this time claiming Fred had pushed an envelope at her, hitting her stomach. She showed no signs of injury, and the magistrate refused to issue a warrant or protective order.

In September, when Judge Hammond's decision came down, she continued the temporary joint custody arrangement, yet again gave Fred physical custody. She also took another action: At Fred's request, Hammond limited Piper's phone calls to the children to between 7:00 and 7:30 P.M. Fred would later say that Piper never lived by the judge's order, calling at all hours, especially after he had put the children to bed. Judge Hammond did one more thing: She hired Dr. Sherman Master, a psychiatrist, to evaluate both Piper and Fred, to advise her on their suitability for raising the children.

That fall, no matter where Fred turned, Piper was there, interjecting herself into the lives of the children. She was at The Jack and Jill School, where Callie was in prekindergarten, eating a snack with the children, when Fred arrived to pick up the five-year-old. When Fred told Callie to come to him, Piper clung to the child. When Piper finally left, she squealed the tires on her van as she pulled from the parking lot as the teachers and children watched.

At Pinchbeck Elementary, Piper showed up often to eat lunch with Paxton, something she'd rarely done before her actions had come under scrutiny of the divorce court. She even signed on as the nine-year-old's room mother. When Jocelyn's Girl Scout troop toured Monument Avenue, Piper volunteered to help chaperone. The evening Fred worked with Paxton on his pinewood derby car in the Cub Scout leader's garage, while Jocelyn and Callie played nearby, Fred looked up and saw Piper in her car in the street, staring at them. He asked her to leave, but Piper remained, watching her children and her soon to be ex-husband, the family that once was hers.

Throughout 2001, Ciulla observed Fred Jablin, wondering how he was bearing up under the strain. She'd heard such bizarre stories about Piper's behavior. She knew he was worried. He talked often of Tina and her influence on Piper, speculating that Tina could be behind some of Piper's behavior. Oddly enough, what she saw was that rather than interfering with his teaching, Fred was becoming more in tune with his students. "All he was going through seemed to humanize him," says Ciulla. "He'd been this rigorous, tough scholar, who demanded the best from his students. Now he was more aware of them as individuals. He understood that sometimes life interfered."

In November, Susanne Shilling deposed Piper on the stand, while Piper's fourth divorce attorney, Robert L. Harris Jr., listened. That afternoon, Piper portrayed herself as an exceptional mother and an exceptional attorney, one who'd done landmark work in school law in Texas. Through personal study, she said she was qualified as a master chef, a mineralogist, a nutritionist, and that she'd taken a correspondence course from England's Cambridge University and earned a designation called Bard, qualifying her as a storyteller and poet. She claimed to be working toward a Druid designation. "The druid level is a wider spectrum of application," she testified. "My name Rountree is an Irish name that means sacred holder of the knowledge of the tree." That day, she talked of angels, and said she'd once written a pamphlet for tennis students she'd entitled "Tennis and the Inner Light." She even claimed to be a master scuba diver who'd once supervised the training of a group of Navy SEALs. When asked if she'd looked for work, Piper said her background "was in raising my children . . . that is full-time employment."

On the stand, Piper's mantra was "for the children." When Shilling inquired about all the credit card bills she'd run up, including purchasing seven tennis rackets, Piper replied, "The kids went through a lot of rackets." When she asked about the $705.40 she'd charged for future services at the Beyond the Fringe hair salon, Piper said, "It was for the children."

Others testified at that hearing, including an employment

specialist who questioned Piper's contention that she'd been unable to find a job. Even as a paralegal, a job for which she was in demand and amply qualified, Piper could have earned up to $40,000 a year.

That fall and winter the e-mails between Fred and Piper were cordial if cool. They concerned Callie's birthday party at Skate Nation, dropping off and picking up the children from visitation, the children's grades, and the distribution of their income tax refund. At times Fred seemed exasperated, trying to pin down Piper on simple issues, like when the kids would be arriving or leaving, or where they'd be staying when she took them to Texas for Christmas that year. In response, Piper chastised him, acting as if he had that information, and saying, "People often get forgetful in times of crisis." She went on to say it frightened the children when he didn't remember things. "PLEASE do not scare the children," she wrote.

Fred forwarded the e-mails to Shilling with his own comment written in the margin: "Very emotional response with numerous fabrications . . . worries me a lot."

Meanwhile, the divorce proceeded.

After the first of the year, January 2002, a flurry of activity began regarding the psychological reports Judge Hammond had ordered, including faxes to the court from the psychiatrist, Dr. Master, and Leigh Hagan, Ph.D., a clinical psychologist Master had asked to consult on the case. When it came time to distribute Hagan's actual report, Piper barred the psychologist from giving it to Fred and his attorney, but Dr. Welton's subpoenaed records from his two years treating Piper went into the divorce record.

On January 7, 2002, nearly a year after Fred had first filed for divorce, Hagan testified in Hammond's courtroom. He recounted his assessments, saying he'd given both Piper and Fred a variety of psychological tests. People who scored like Piper did, Hagan said, tended to be suspicious and place blame on others, they had difficulty with personal account-ability and were quick to anger. In Hagan's assessment, Piper was recklessly impulsive and had difficulty dealing with change. "They find loopholes and work around them when they feel corralled by somebody else's regulations."

That day on the stand, Hagan speculated on five possible diagnoses for Piper: ADD, bipolar disorder, disassociative disorder (in the past called multiple-personality disorder), substance abuse, or a character disorder. All suggested that Piper might be hyperemotional and have impaired judg-ment. "I found a greater weight of a theory of disassociative disorder," Hagan said, pegging a likely reason as Piper's fa-ther's alcoholism and stroke.

"She was not able to rule out for me the possibility of hav-ing been sexually victimized in her childhood," he testified.

When Hagan talked of Fred, he had nothing damaging to report. In fact, he said Fred had high moral aspirations and that people who answered as he did on the tests tended to "go along to get along. Tend to be peacemakers and tranquil, and didn't have a hair trigger temper." Fred, he said, was a man who would rarely anger and would rely on tact and di-plomacy, not being "a bull in a china shop."

On the stand, Dr. Master would disagree with Hagan about Piper, discounting a possible diagnosis of disassociative dis-order, which could result in periods of forgetfulness and feel-ings of being "outside of themselves." He labeled Hagan's assessment as "a theory, that's all." Perhaps Piper didn't tell Master what she'd tell others: that at times she lay in bed and imagined transporting herself out of her body, across the miles, into her children's bedrooms while they slept.

"She said she whispered in their ears that she loved them," says a friend.

Contrary to Hagan's assessment, Master's diagnosis for Piper was ADD and an adjustment disorder that mixed anx-iety and depression. He blamed it on her limited access to her children. "I don't know that her disorder would cause her to be incompetent as a mother," he testified. When questioned by Shilling, Master said that Piper had been on stimulants for her ADD, as well as antianxiety and antidepression medication, since 1994. Under questioning by Berkeley, the guardian *ad litem* appointed by the judge to represent the children's best interests, Master testified that to manage Piper's condition would take long-term therapy.

"If she got consistent help, she would be all right," he said.

When it came to Fred, Dr. Master said that Piper's husband saw himself as blameless in the conflict, although "it takes two to tango." And he speculated that Fred could, under stress, overreact.

Something else came out that day under questioning by the judge: Master said that Piper had told him she'd begun the affair with Gable in February 2001, before the divorce was filed. "That is a specific date she gave me and she even put it in writing," he said. That date was important, since Fred's grounds for the divorce were adultery.

On the stand that day, Ana, the children's nanny, also testified. She recounted a day Piper had broken a window to get into the house. Ana walked in and found Piper washing her clothes in the washing machine and cleaning up the glass. "I don't want the kids to see it," she told Ana, who'd called Fred immediately to report the break-in.

"The children already know," Ana told Piper, describing her as agitated.

Also before the judge, Fred's attorney placed letters from the children's therapists, dance and music teachers, and athletic coaches, documenting that Piper didn't take them to their activities, or even counseling sessions intended to help them through the emotional barrage of the divorce. Perhaps the saddest moment was when Jocelyn's therapist testified how deeply depressed the twelve-year-old was.

On the stand, Dr. Gable, Piper's lover, testified on her behalf. She was a good mother, he said, and he described Fred as an absent father. The physician, appearing nervous and wringing his hands, contended that Fred put undue pressure on the children, citing a day Gable and Piper were taking Jocelyn to an art lecture. Fred had refused to let her go, wanting her to study because she'd failed a math test. Jocelyn had sobbed, crying, "What do I do?" over and over.

When Shilling took over the questioning, another picture of the physician emerged, as she listed events the children had missed while they were allegedly with him and Piper. The doctor insisted he knew nothing of the children missing doctors' appointments and soccer practice. Under questioning, he did, however, admit kissing Piper in front of the children. And there was something else, something many had speculated about for months. "Finally, sir, let me ask you this, can you tell me why you prescribed drugs for Ms. Rountree?" Shilling asked.

After a pause, he answered: "I did it on a few occasions, as a favor."

"I have learned what my children like to eat for breakfast," Fred began when he testified at a February custody hearing. He talked of waiting at the school bus stop and taking Callie to preschool, playtime and bedtime reading, tucking them into bed and kissing them good-night. This was a new Fred Jablin, one who baked a cake for Paxton's class and bought Jocelyn's first bra. For the judge, Fred described each child: Joce and Callie were happy children, doing well, while Paxton, who'd been closest to his mother, was having the worst time of the divorce.

The children, he said, were suffering from the chaos Piper brought into their lives. He talked of the Sunday night he was cooking dinner after picking the children up at Piper's, when Joce handed him a note. "Why are you holding us hostage?" it read. He didn't believe she'd written it willingly, but that Piper had dictated it to her.

That night, Fred sat with the children, trying to explain the intricacies of divorce. He wasn't the one who would decide if they lived with their mother or with him, he said, that decision rested with the judge. The children cried, upset and confused.

Yet, despite all they'd been through, Fred insisted he didn't want to push Piper out of the children's lives. "I value the time my wife spends with my children. I have a tremendous amount of compassion for my wife, and I know that she loves my children very much," he testified. That understood, he said he wanted Piper to abide by the rules, not to interfere with their lives. Fred insisted he'd been functioning as a single parent for a long time, and that Piper was periodically so dysfunctional that she sat and literally stared at the walls.

That wasn't all, he said. His ex-wife frightened him.

Piper glared at him when he looked up and saw her at Paxton's soccer games. And there was something else, something he found very troubling: Piper had taken the children to Richmond's historic Hollywood Cemetery, a gracious, rolling landscape with views of the James River and the downtown skyline. It wasn't where they'd gone that bothered him. The cemetery was a popular tourist destination and a beautiful place, where history came alive. There, U.S. Presidents James Monroe and John Tyler, and Confederate president Jefferson Davis are all buried, along with 18,000 Confederate soldiers.

It was what Piper had done with the children while they were there that Fred found distressing.

Amidst the stained, aging headstones, the sculptures of angels that looked toward the heavens, and solitary women who leaned forlornly against crosses, Piper had posed the three children for ghostly photographs, their eyes blackened and their faces solemn.

"It was frightening," Fred told the judge.

Others testified that day, including Tina, who portrayed Piper as a good mother. She described Fred, on the other hand, as insulting to the children, highly critical and hurtful. As an illustration, Tina said Fred had once called Joce "stupid" for spilling milk. Rather than Piper, she said Fred suffered from depression, and she testified under oath to what her ex-husband, Dr. Howard Praver, would later label as a lie: that he gave Fred a prescription for Prozac.

On cross exam, Shilling suggested Tina's testimony was a "fabrication."

All was not going well for Piper's side that day in the courtroom, which should have been obvious when the judge told her attorney: "Your client's credibility is an issue, because I have three columns of things here that she told me at different times that are not consistent." Still, the following month Piper submitted her suggested custody plan to the court, a joint arrangement where she and Fred alternated weeks.

At a convention that year, Fred sat with Daly, Miller, and other colleagues, old friends all, and told them of the tragedy of his marriage, the slow, painful unraveling that was consuming his life, costly both in attorney's fees and emotional aching. "How could a mother do these things to her own children?" he asked. "Why would she?"

Although Fred found no real answers to his questions, which he'd repeat over and over throughout the ordeal, in the courts the case was nearing a conclusion, at least regarding custody.

In a letter dated March 5, 2002, Judge Hammond mailed her decision to all those involved. Both parties, she said, loved their children, but the animosity between them was potentially detrimental to the children. "Dr. Jablin is more of a peacemaker, more able to bend," she wrote. In the letter, the judge singled out the testimony of Jocelyn's psychiatrist, who'd testified that she'd improved during the time Fred had custody. For those reasons and others, Hammond ruled that while Piper and Fred would continue to share joint legal custody, Fred would have permanent physical custody of all three children.

When the decision came down, Piper was inconsolable. She cried on the telephone with friends. When they asked why the decision had gone against her, she didn't talk of her adultery or her bizarre behavior, but of Fred. He was a master of manipulation, she said, and he'd connived to foil her in the courtroom, using his mastery in communication. But for her it all boiled down to one defeat:

"I'm going to lose my babies," she told Annie Williams.

Three days after the judge's decision came down, Piper retaliated.

Although Tina was not a psychiatrist or a psychologist, the previous November she had written a psychological profile of Fred. The result was far from flattering.

Now, with the judge's decision against her, Piper e-mailed it to friends, family, and neighbors. Attached to an e-mail from Piper's piper_lightworkergalleries@yahoo.com account, the report was referenced as "Jablin Psychological Profile" and marked: "Relevant if you have children he supervises." Although the forty-two-page treatise had nothing to do with the court case, Piper introduced it by writing: "After serious consideration of the wider implications of the attached *court report,* I am forwarding it to you."

From page one, Tina got right to the point: "I am writing this letter out of grave concern for the current and future welfare of the children who have been placed under the care of Fred Jablin." She began by presenting her credentials. Some of it was blatantly untrue, as when she described her clinic as in the "heart of the Texas Medical Center." In truth, the clinic wasn't located within the acres of skyscrapers and hospitals that comprised the Medical Center at all, much less at its center. Tina also claimed to hold an honorary position as a professor at the University of Texas for her "expertise in pediatrics and women's health issues." Later, a UT spokesperson would find no record of Tina having ever filled any such position at the school.

From that point on the paper was a direct assault on Fred Jablin's character, morals, and ethics. Tina labeled Fred a narcissist, who cared about no one but himself, and said that he had no empathy for others, including Piper or their children. "I am HIGHLY ALARMED that the children, Jocelyn, Paxton, and Callie have been turned over to Fred," she wrote. She accused him of "violent and dangerous outbursts" and a "calculated pattern of intentional neglect." He should never have influence over small children, she wrote, underlining the passage for emphasis. She claimed Fred regularly smoked marijuana and had physically and sexually assaulted Piper, and deprived her of money and affection. Tina contended that Fred suffered from a "pervasive pattern of grandiosity . . . a need for admiration and a lack of empathy." When they were with him, she said, the children were "at a high risk for psychological abuse."

"Fred Jablin actively oppresses any student [sic] (or child's) needs which are contrary to his viewpoint," she wrote.

The headings shouted out: "Suicide and Dr. Jablin's Psychological Manipulation of his Children," and "Expectations of 'Entitlement' of the Narcissistic Personality Disorder."

Fred had been so excited about Jocelyn's birth he'd shown the video to his friends, yet Tina wrote that he had only disdain for children and had never wanted them. And she again made the claim she'd made in the courtroom, one that her ex-husband would vehemently deny: that Dr. Praver treated Fred. "After seeing for himself several alarming incidents of

Fred's rage and stress reaction when he felt out of control, Howard discussed with him the merits of dealing with his stress with the medication Prozac. Thereafter my husband wrote him a prescription for Prozac for years."

Tina went on to assert, despite Jocelyn's psychiatrist testifying to the contrary, that in Fred's care "Jocelyn has been bordering on anorexia and suicide for months." She said that Fred had beaten Paxton, and she even took a swipe at Fred's religion, saying that Jews had a persecution complex.

After it arrived on her computer, Melody forwarded the e-mail to Fred. It was a blind e-mail, and there was no way to tell who else had received it. It hit Fred hard. He called friends and colleagues, anyone he thought might have it, to try to explain that the report wasn't a court document. "It's a miserable creation of her sister, Tina Rountree," Fred e-mailed to Paxton's Cub Scout leader, one of the many who received it. One recipient was connected to the University of Richmond. That week, Fred withdrew his name from consideration for the deanship of the leadership school.

In hindsight, that e-mail would have a chilling effect on what was already a deteriorating situation.

With no idea who had or hadn't received the e-mail, Fred acted as if anyone he met, anyone he dealt with, could have read Tina's damning psychobabble. He cornered people at work, at the children's schools, friends in the grocery store, and neighbors on the street to ask if they'd received anything odd from Piper that pretended to be a court document. At a school event, he approached one parent, asking her if she'd received the e-mail, trying to explain. She looked at him as if he'd lost his mind, unable to understand why Fred would appear so out of control.

Two days after receiving the e-mail, Fred sent one to Piper: "The defaming messages and attachments you have been broadcasting via e-mail and attachments are detrimental to the children's welfare." Angry and hurt, he then rescinded an invitation he'd extended to Piper to come to the Hearthglow house for Paxton's birthday party that weekend. On the thirteenth, he took more serious action, filing a formal petition with Judge Hammond for an emergency hearing on custody, asking for fewer visitations for Piper, because of the "psychological problems underlying the defendant's conduct."

That spring, the Jepson School had a new dean, Ken Ruscio, who'd come from Washington and Lee University, where he'd been a professor in the School of Commerce, Economics, and Politics. Despite what must have been his bitter disappointment, Fred was helpful, guiding Ruscio through the transition. Perhaps it was all for the best, he told a friend. Now, after all that had happened, he knew he needed more time for the children. Increasingly, the children were becoming Fred's world.

Still, there was no way to completely end the struggle, to excise Piper from his life. After all, they had been married for nearly two decades, and they had the three children, a bond that Fred could never completely break.

In May, Piper showed up at Byrd Middle School and demanded that Jocelyn be released to leave with her. She was furious when the school, following the court orders they'd been supplied, refused. That afternoon, when she was supposed to pick up Jocelyn for visitation, Piper showed up forty minutes late and then, again, didn't take her to ballet class.

When school let out for the summer, Piper had her first job since the divorce, as an art counselor at a Girl Scout camp. In e-mails she offered to bring the children with her. Despite all that had happened, it would later seem that Fred was still willing to include her in their children's lives. He agreed, but before it would take place, Piper had quit.

The push and pull of visitation went on throughout the summer, as they negotiated how and when the children would see their mother. Everything became a battle of the e-mails, from setting up summer tutoring to the children's report cards. Although he didn't need her permission, Fred consistently advised Piper of his plans; uniformly, she disagreed, as if attempting to retain some control of her children's lives. When Fred asked her for information, Piper simply didn't answer, leaving him and the children uncertain.

By then many on Hearthglow Lane were noticing the toll the divorce was taking on the Jablin children. They'd become quieter, more reticent to play with friends, withdrawn. "Our hearts just went out to them," says a neighbor. "We knew all three were suffering."

On July 15, Fred and Piper were again in court, this time to discuss her offensive e-mail. In his petition, Fred had asked the judge to order her to give him the names of all those she'd e-mailed it to and to ban her from sending it to anyone else. At the hearing, Piper, who'd gone through five attorneys, represented herself. "All through the divorce, Piper had been acting crazy with Fred, then showing up in the courtroom controlled and acting like a lawyer," says Melody. "Once Piper sent that e-mail, the judge had an indication of what was really going on."

The divorce and custody still weren't final, and Fred wanted the judge to rule an end to what had been more than a year-long court battle.

On the stand that day, Shilling walked Fred through the e-mail, asking if the allegations of violence and psychological abuses were true. "No," he answered. What bothered him as much as anything were the stereotypes in the e-mail about Jews. "It says we have a hereditary predisposition, a genetic predisposition, or some type of complex, or whatever that her hack sister . . . perpetuated, and this woman is spreading. My children are, in fact, part Jewish, your honor." As he said "this woman," Fred pointed at Piper in disgust.

That Jocelyn was anorexic or suicidal flew in the face of what her therapist had testified to in front of the judge. Fred pointed out, "He said she was doing well."

"I object," Piper said.

"She is not anorexic," Fred shouted, characterizing the e-mail as "close to a hate crime."

That day a strange event would take place, as Piper, acting as her own attorney, questioned Fred on the stand. As if not realizing everything she'd put him through over the past two years, she produced the loving e-mail he'd sent her soon after they'd separated, in which he begged her for another chance to save their marriage and family.

"You still loved me. What changed?" she asked.

Fred answered, "I am no longer willing to stay together for the sake of the children . . . you've been involved in an adulterous relationship . . . for an extended period of time."

"So you think that because of these incidents you're alleging, my judgment regarding the children is impaired?"

"Yes," he answered.

With that, Piper attacked him about such matters as the children no longer taking piano lessons. He countered that they couldn't, that she'd moved the piano out of the house.

"Have you asked if they want to continue tennis lessons?" she asked.

"You took all their tennis rackets," he answered.

That day on the witness stand, with Shilling asking the questions, Piper repeated her mantra, invoking the phrase "for the benefit of the children" and saying "the children need their mother."

"Specific provisions need to be provided to ensure that the children have access to me as much as possible," Piper said. When the judge said distributing the e-mail had not been in the best interest of the Jablin children, Piper replied:

"I have been terrified for myself and for the sake of the children. My concern is the welfare of the children."

With that, Susanne Shilling sat down and asked no more questions.

Tina was there supporting Piper, as she had throughout the divorce. They were two sisters fighting the system and one soon to be ex-husband, whom they both passionately hated.

"Did Piper condone the report?" Shilling asked.

"It's my report," Tina answered.

"Are you saying you feel you knew more than the court knew when it made the order?"

"Absolutely."

With that, Piper asked for an increase in support payments. Although she wasn't the one supporting the children, Fred was, she pleaded, "The children and I are in desperate need of help and support . . . the children need to be given a chance to be heard."

As that painful day in court drew to a close, Shilling, on Fred's behalf, asked the judge to end the drama with a final ruling of divorce. The issue of money still remained, awaiting a report from the court-appointed commissioner on the best way to divide assets. But that would wait for another day. This day, Judge Hammond did as Fred requested: "You've got it. I've signed the order," the judge said. "You're divorced."

On the stand, Fred had made another request—that the joint custody decree be vacated and that he be granted full permanent custody of the children.

"I cannot trust Piper's judgment," he said. "This e-mail shows a person who does not know right from wrong, has no moral standards, will lie and spread hatred. She does not have the welfare of the children at heart."

Before adjourning, Judge Hammond issued one more ruling: She limited Piper's visitation to the children to twice-monthly weekends, holidays, and three weeks every summer. And she granted Fred's petition, awarding him sole custody of all three children.

Not long after, Piper told a friend: "Fred torched my village."

9

Piper hated Fred for taking the children away," says a friend. "She saw herself as the victim, and he was punishing her for the affair."

Bitter custody battles forever wound parents and children. For many mothers and fathers, losing custody is akin to a death, a loss so raw it tears a hole in the heart. For Piper Rountree, a woman who based her self-worth on being an exceptional mother, it cut so deep that friends feared she could never recover. Piper blamed everyone but herself for the decision: Fred, the judge, her lawyers, and the Virginia legal system, contending the state didn't "protect mothers and children."

Even those who believed Piper saw the children more as her reflections than as individuals with their own wants and needs wouldn't deny that she'd made Jocelyn, Paxton, and Callie the center of her world. "I am a mother first, everything else second," Piper would later say. Yet, it was also easy to see that she used the children as an excuse for her failures. At one point Judge Hammond had asked why she'd quit her job with the Girl Scouts, the only one Piper had held since moving out of Hearthglow. Even though the children lived with Fred not her, she answered: "Taking care of our children is a full-time job."

There was also another matter that made the judge's decision a disappointment for Piper: If she'd won custody, Fred would have had to pay her child support. Now that he had permanent custody, that prospect evaporated. In little more than a year her entire world had changed. She was still seeing Dr. Gable, but she told friends that the relationship had cooled. She hadn't truly worked since leaving Texas, and, with nannies and cleaning ladies, for the past eight years she'd been free to play tennis, paint, hike, and rollerblade with the children, and drink wine and relax with friends. Now the children were rarely with her and she had to pay her own bills. Working wasn't something Piper had a natural inclination toward. Later, her contention that she'd applied for scores of jobs would seem questionable. "Piper could have found a job in Richmond," says Mel. "She just didn't want one."

Having used up the $7,000 in prepaid moving expenses circulating from house to apartment to house in Richmond, and citing her inability to find work, a month after the custody decision Piper packed her possessions in a rented trailer and attached it to the back of her 1994 Chrysler Voyager to move to Houston. Steve Byrum, a tall, rugged-looking man with a dark beard, who'd met her through their sons' Cub Scout troop, helped her load the furniture and boxes. Curious about what had happened in the divorce, at one point he came right out and asked: "Why didn't you get custody of the kids?"

Piper's eyes filled with hate. "Because Fred's an asshole," she hissed.

As usual, Piper hadn't kept Fred informed of her plans. He learned of the move through the children, and he expressed misgivings, both about the children traveling to Texas for visitations and Piper being so close to Tina. "They feed on each other," Fred told Ciulla at work one day. Then he commented that the past two years had left him feeling trapped in a Kafkaesque novel, in which his wife had morphed into a troubled, vengeful woman: "She's just not the woman I married."

Days later, Fred filed a complaint with the court, charging that Piper had violated court orders by not informing him of her plans and where she'd have the children during visitations. By then, in Texas, Piper didn't show up at the hearing.

In Houston, Piper moved her possessions into storage, then stayed briefly at Terry Wichelhaus's house, to care for her children for a week while Terry was out of town. Terry, whose house was decorated year-round with all manner of fairies, says that with Piper's petite athletic body, she reminded her of Peter Pan: "She was almost fairy-like. Her face just lit up when you talked with her."

Still, Terry thought it unusual when one of her friends cornered her after she'd spent time with Piper. "Did you know that Piper believes she's a Druid priestess?" the friend asked.

In Irish and Welsh legends, Druids were sorcerers and prophets, members of an ancient Celtic society of priests who worshipped the moon and the natural world. They made potions and believed in reincarnation. Perhaps Piper felt a kinship with the Druids because legend said they carried staffs of rowan wood, the tree that gave the Rountree family its name. The revelation of Piper's belief was news to Terry, but not shocking. "Piper thinks she has magical powers of some sort, and Tina thinks Piper has magical powers of some sort," Terry would say later.

When it came to Fred, Piper left no doubt how she felt about him. "She hated Fred for taking the children and for trying to prove she was crazy," says Terry. "Her family hated him, too. It was a huge topic of conversation; everyone in the Rountree family was talking about how Fred had screwed Piper. Drug addicts and prostitutes lose their kids, not good moms like Piper."

Something that happened the week Piper babysat, however, changed Terry's view of Piper as a mom. As with her own children, Piper was a playmate to Terry's, rollerblading and going out for ice cream, acting like a kid herself. But then, while babysitting, Piper brought a young doctor she'd met to Terry's house, and he stayed overnight. "I never would have let her stay with my kids again," says Terry. "She should have known better."

After the week at Terry's, Piper moved into Tina's gray Victorian corner cottage with the graceful porch overlooking the street. As the months passed, Tina would try to comfort Piper, who cried, yearning for her children. At times, she was nearly inconsolable. With anyone willing to listen, Tina made no secret of her disdain for her now ex-brother-in-law. Fred, she said, her voice thick with anger, was manipulative and evil, abusive and cruel. "Tina absolutely hated Fred's guts," says a friend. "She held him in complete contempt."

At the time, Tina's boyfriend, Claude "Mac" McClennahan, a six-foot, blue-eyed disc jockey, lived with her. Since the seventies, Mac, who'd grown up in the Texas Hill Country, had worked on radio stations throughout central Texas and Houston, mostly playing rock and roll. In his mid-forties, he looked the part of an aging rocker, plump, with a gray-streaked beard and mustache, and a receding hairline that fanned out into shoulder-length dark brown hair.

Before becoming a DJ, Mac had worked as a cook and a restaurant manager, once even steam-cleaning egg trucks. He played drums in a band, and had a penchant for tennis shoes and rumpled jeans. "I'm a generally nice guy," he's said, and others thought of him that way as well. A computer junkie, Mac linked up with Tina in 2000, through the Internet dating site Match.com. In some ways, friends say he brought structure to her life. At the time he was between radio jobs, and he took over the management of the clinic, overseeing the business end she hated. Years later Tina would describe Mac as "a nurturing kind of guy who watched out for me." When it came to Piper, Mac would say he grew to see her as his "little sister."

From that point on, Piper had entered Tina's world. With patients and friends, Tina had a reputation for offering guidance on how to live their lives. She would do nothing less for her little sister.

Even before Piper drove the U-Haul into the Houston city limits, Tina approached Martin "Marty" McVey, a fiftyish attorney who lived on the top floor and officed on the first floor of a duplex just doors from the clinic. A former Houston prosecutor, McVey, white-haired, bearded, and with a generous girth, played Santa at a Houston children's hospital every Christmas. A former high school and college football player with a fondness for cowboy boots, he says, "I'm not a pin-striper. I'm a country boy, and in a courtroom I down-home people to death."

Years earlier McVey had rented offices from Tina above her clinic. Still, he maintained that they weren't close friends. Instead, McVey was a friend of Tina's second ex-husband, Dr. Praver, and contended that he had mixed feelings about her, saying she'd taken advantage of Praver. "Tina is a user," he'd say. "She uses everyone around her for her own personal reasons."

Tina wanted McVey to assist Piper, by taking her under his lawyerly wing, renting her an office, and helping her re-start her Texas legal career. McVey, who works alone in a front office, had an empty back office he didn't use and agreed to rent to Piper for $500 a month.

At first the arrangement went well. Piper moved in and set up a desk with a computer, bringing in the children's artwork to decorate the walls. Tina dropped in often, and McVey had time to see the sisters together. As others had over the years, he came to the conclusion that Tina dominated Piper. "When Tina was around, Piper was submissive," says McVey. "Tina was always telling Piper what to do, and Piper was doing it."

When Tina wasn't around, Piper was bubbly and fun. Yet, often all she wanted to talk about was her divorce and how she'd been abused by the system. She had records from her case in binders and spent hours combing through them, searching for a way to restart the fight to get the children. At times she asked McVey, who did criminal and divorce work, for advice. He deferred, saying the laws in Virginia were different than Texas. But he found it hard to understand how she'd failed to get the children, even if she'd committed adultery. "The courts nearly always give the kids to the woman," says McVey, "unless something's really wrong."

The children came to visit, and McVey watched Piper with them. The living room/reception area, homey with a fireplace, hardwood floors, and burgundy leather furniture, became their art area, where Piper pulled out construction paper, paints, and scissors. More than once McVey returned to the office from court and found the office windows and walls plastered with the children's art, and Piper and the three children all on the floor painting.

"She was so into those children," he says. "More than any mother I've seen."

It would later seem incongruous that in the fall of 2002, after her own bitter divorce ended in her losing custody, Piper took a seminar that, with her Texas law license, qualified her to work as a guardian *ad litem*, an attorney charged with overseeing the best interests of other people's minor children in divorce cases. Before long she was being appointed to work cases by judges. McVey, too, handed her work, small cases he didn't have time for or ones that didn't interest him. In return, she agreed to pay him a percentage of her fees when the cases were settled. Piper seemed happy for the work and grateful for the chance of an income. But after the first few months, each time McVey asked Piper about the rent, she replied, "I don't have it." He thought about telling her to move out, but a woman he was dating convinced him to let Piper stay.

"She's had a rough time," the woman said. "Give her some time to get her life together."

Away from the office, however, McVey had the sense that Piper was living life on a grand scale. She dressed well, from expensive business suits to cowboy boots and jeans. Piper and Tina frequented the restaurants and bars in the area and rubbed elbows—using the membership Tina had gotten in her split from Grant Heatzig—with the upscale crowd that frequented the Houstonian Hotel health club. "Tina's always lived well . . . and Piper moved right into that world," says McVey.

In November, Terry, Tina, and Piper went to a popular Houston bar, Sam's Boat, a busy, noisy place where the crowd is young and there to meet others. "Whenever you went anywhere with Piper, the men went crazy," says Terry. "She was flitting from lap to lap, flirting. Two or three men asked Piper for her phone number that night. She just had that effect on men."

One of the men was Jerry Walters, a six-foot-one, fifty-year-old oilman from Baton Rouge, Louisiana, with a well-lined face, prominent chin, a deep whiskey voice, soulful brown eyes, and a penchant for big silver belt buckles, snake-skin cowboy boots, and flirting. "You gotta help me find a honey. You know I'm a lonely man," he'd say with a crooked smile. In truth, Walters needed no one to help him find women. Divorced for years, with one grown daughter, he worked out daily at a gym and had the massive chest and legs of a body builder.

Walters and Piper began as friends, but by April 2003 they were lovers. Piper seemed fascinated with the big, bulky oilman with the hearty laugh. If her first lover after the divorce, Gable, looked enough like Fred to have been his brother,

Walters was both their polar opposites. Monday through Thursday, Walters worked in Houston, while on weekends he drove home to Baton Rouge and his beloved bloodhound, Bertha, and his "main woman," his mother, who suffered from Alzheimer's. His connection with Piper, he'd say, was strong. She was different than any woman he'd known. It was the way she could spend an hour simply watching a squirrel forage in a tree. "She enjoyed life more than anyone I've ever known. Just sipping coffee in bed in the morning and talking was an event when I was with Piper," he says. "She lived totally in the moment."

At times, Piper talked about her divorce, and "poor Fred, whose wife was unfaithful."

Still, it was another world that seemed to call to Piper. The business cards she had printed would perhaps attest to what was going through her mind that fall. On the front, she superimposed her own photo, serious and arms crossed, in front of a black and gold painting of a classic Greek figure, a woman warrior wearing a headdress and brandishing a sword. Printed over the photo was: "The Law Firm: Piper A. Rountree." Inside, the card touted her talents as a former district attorney and a school attorney. She claimed to have twenty years of experience, and wrote: "If you feel you don't have the Right . . . or are convinced that you are not deserving . . . then we need to talk . . . If a person puts up with whatever bad situation or injustice is imposed upon them, they are really just teaching their children to learn how to suffer as well."

Perhaps that's what she thought she would do if she accepted the loss of custody—teach her children to suffer. On the back of the card was one of Piper's paintings, a dark angel on a fiery background, holding scales and a sword.

Not all was calm at Tina's house that year. The relationship between Tina and Mac had always been volatile. Later, she'd say that over the three years they were together, he walked out on her three times. In 2002, not long after Piper arrived, Mac was getting ready to leave and apparently worried about how Tina might react. One day he walked up to Piper and handed her a .38 caliber revolver Tina kept in the house. For some reason, perhaps his history with her, Mac worried about having it in the house when he told Tina he was leaving.

"He asked me to get rid of it," says Piper. She did as requested, entrusting the gun to McVey for safe keeping. Days later Mac moved out, but as he had in the past, he later returned to Tina's house and her life.

Meanwhile, in Richmond, the Rountree-Jablin divorce continued, but the battlefield had moved from custody to money.

In September, Commissioner Bischoff held a hearing that had been scheduled the preceding summer. Piper had been in the courtroom when the date for the hearing was set, but she didn't show up, instead sending an attorney to ask for a continuance. Saying Piper had been given ample notice, Bischoff refused. Since his client wasn't there, Piper's attorney left. The hearing went on as scheduled, without Piper or an attorney present to represent her.

That day, Susanne Shilling and Fred laid out for the commissioner, who'd be deciding the financial settlement, the horrors of his years with Piper, including her adultery, flagrant spending, impulsivity, and her false charges of domestic abuse and neglect of the children. Shilling asked Fred, "Did you love her up until the time you actually separated?"

"Yes," Fred answered. But he added that after the split, "I always thought there was the potential for Piper to do very destructive things to me."

Shilling then turned her attention to the matter at hand, the financial settlement. Fred and Piper had property: the houses in Austin and on Hearthglow, the beach house, and retirement accounts. Piper had little in her retirement account, but Fred had more than $300,000 in 401(k)s at UT and UR. But he'd also taken over $70,000 of the family debt, much of it from credit card bills attributable to Piper, and his salary had been cut now that he was no longer acting dean. The house in Austin had been his before the marriage, and the beach house was purchased using separate property— his inheritance from his parents. By the time Fred paid the bills, including a nanny for the children, and despite grossing nearly $10,000 a month, he claimed a negative monthly cash flow of more than $3,000.

At the hearing, Fred read from a letter Piper had written to Callie: "I'm pretty sure I have a job with a large law firm," she said. "I'll know this week for sure."

"She has the potential to make more than I do," Fred testified.

Afterward, it would seem odd that Piper, an attorney, had misjudged the situation so badly, by not showing up at a hearing where her financial future was at risk. When the decision came down, it was devastating. Judge Hammond, on Bischoff's recommendation, not only discontinued the support money Fred had been paying Piper, but ordered that she begin paying Fred $890.24 monthly for child support.

That support order would later appear to have gotten Piper's attention.

In December she wrote a letter to her most recent attorney in which she listed twenty points she said should be used to vacate the order, including her contention that the ruling wasn't within the judge's power.

While Piper stewed over her increasing misfortune, nearly fourteen hundred miles away, life continued for Fred and the children. They spent that Christmas at the beach house. Later, he e-mailed Piper that they'd had a traditional celebration, with a big dinner, a tree, stockings, Santa, and gifts. "We had fun, and the spirit of Christmas was in the air," Fred wrote. In her reply, Piper said acerbically, "Thank you for having some X-mas for the kids in spite of your undertaking of otherwise following Jewish principles."

Her own Christmas was less successful: dinner with Tina and friends and a trip to the hospital when Tina's youngest son cut his hand on a broken window.

"I miss my babies so much it hurts," Piper cried when she called Loni Elwell.

Throughout that year and into the next, 2003, Piper and Fred corresponded via e-mail. He wrote her of the daily joys of life with the children, celebrating the day Jocelyn's braces were removed and when she made the honor roll, talking of Paxton's and Callie's activities and their good grades. For the most part, Piper was civil with him, even thanking him "for letting me know what is going on with the children." But just below the surface the power struggle continued. The court hadn't divided Fred's pension funds yet, and Piper wanted him to pay to fly the children to Texas for her visitation. He refused.

"I have my own problems," he wrote. "With my responsibilities and the debt I am saddled with."

To save the three airfares, at times Piper went to Virginia, checked into the TownePlace Suites hotel, and had the children stay with her there, just as Fred had when she had a restraining order against him and he was barred from the house.

In January, Piper moved out of Tina's and into an aging wood-sided house a block from McVey's office, with an $850 monthly rent. On a quiet street, set back from the road, backing up to a greenbelt, it had space heaters and window air conditioners, a long way from the comfortable homes she'd lived in with Fred. The night Vincent Friedewald moved into the old duplex next door, he saw Piper gardening well after dark, planting a patch of flowers near the front steps. Wearing a tank top and short shorts, she looked attractive, and she invited him in. Offering him a beer, she had her cell phone tucked into her bra, and it vibrated and chirped like a cricket when it rang, causing her to erupt in laughter. When he got ready to leave, she invited him to "come by and have a drink anytime. The door is always unlocked." He never took her up on it, and he had no doubt she was offering more than a cold beer.

That spring, Piper complained constantly about her lack of funds.

As always, she blamed her situation on everyone else. "You know, I have always had at least a receptionist, one or two secretaries, a runner/filing clerk and law clerks, data entry person, accountant, and a house computer systems person and librarian, practice manager," she e-mailed Fred. "I mention the above because it is very hard for me to function as an attorney when I am barely surviving. I am cleaning my own office."

In Virginia, the children were beginning soccer, and Fred reported on their teams to Piper in e-mails. In response, Piper asked Fred to send her the children's schoolwork, art, and papers. The separation was difficult, and these artifacts had become precious to her. Despite all he'd been through with her, Fred did as she requested, and when Jerry Walters gave her an airline ticket to Richmond to visit the children in March, Fred agreed to let them stay with her at the hotel, even though she didn't give the court-ordered two-week notice. In the e-mail where he agreed, Fred pointed out that the children did, however, have activities that weekend, and he quoted from the court order that stipulated visitation shouldn't interfere.

Finally, in Houston that spring, Piper filed for bankruptcy. On the forms, she listed total debts of $289,596. She owed Tina $15,000, her mother $80,000, and $76,000 on credit cards—over and above the debt Fred had taken on. Later, Fred would argue that many of the debts were fabricated, including the $100,000 she listed as owing him. For monthly expenses she listed $6,390.

Although child support isn't dischargeable in a bankruptcy, the filing had the potential of stalling the Virginia divorce case, as the federal bankruptcy court took jurisdiction over Piper's finances. To prevent that from happening, Fred hired a Houston attorney, Don Knabeschuh, and filed yet another petition, this one asking the federal court to deny the bankruptcy. Knabeschuh would later recall the way Fred unemotionally described the turbulent events of the past years, as if he'd repeated them so often to so many that he'd grown immune to the pain. At Piper's Houston bankruptcy hearing, Knabeschuh attended in Fred's place. When he explained to the judge that Fred was his client, Piper glowered at him.

In Virginia, the Jablin children's dog died, run over in the street after it ran out of the gate. The children were sad, and Fred packed them up for a weekend at the beach house. In Texas, Piper didn't show up at a second bankruptcy hearing and failed to file the required reports and schedules. The judge dismissed the case. Piper filed a second bankruptcy petition, and again Fred protested through Knabeschuh. This time, when Piper met with Joseph Hill, a bankruptcy trustee appointed by the court, she tried to convince him to use the bankruptcy court to set aside the Virginia divorce settlement. When Hill told the judge of the conversation, the judge terminated Piper's second bankruptcy with prejudice, barring her from filing a third time.

That summer, Piper didn't attend her twenty-fifth class reunion in Harlingen, but gossip swirled about her, news of her divorce and the scandal of losing her children, her old friends comparing notes like high school girls gossiping about who'd disappeared with a boy into a dark corner at

Saturday night's dance. If there was a consensus, it was that Piper's ex-husband must have manipulated the judge. How else could it be explained? To her hometown crowd, Piper had always been an "A-lister," brilliant and beautiful. It was hard to understand how her life could have taken such a tragic turn.

Off for the summer, Fred offered to either send the children to summer camp or take them to New York, to see where he grew up. The children chose New York, and he was thrilled. While there, he took them to a Broadway play, as his mother had once done with him.

In Kingsley, neighbors had watched Fred change. He'd always been viewed as a good dad, one who played with his children, but now Fred had become "Mr. Mom," a dad who chauffeured his three children to soccer, planned birthday parties, and called moms to set up sleepovers. During the school year, he joked with the mothers and children on the corner as they all waited together for the yellow school bus to arrive. As the bus pulled away, he'd wave at Paxton, who would wave back with a smile. One evening Fred watched *Lord of the Rings* with his son. As he did in the classroom at the University of Richmond, Fred used the movie's characters and their actions and words to illustrate methods of leadership. "Paxton couldn't believe they watched a movie like that in college," says a neighbor. "He thought college must be cool."

Meanwhile, Piper continued to argue with Fred via e-mail, demanding he pay to send the children to her for visitation. She claimed the judge had ordered him to do so, which was blatantly untrue. At times it seemed that she tried to bully him into doing what she wanted. This time Fred stood firm, refusing, and wrote her, "Your continued delusional assertions and fabrications trouble me greatly." The tenor of their correspondence became even more heated, with Piper claiming Fred had forced her into bankruptcy. He labeled her claims were "pure lies." He also wrote that he knew she'd hit Callie and given her "dubious" prescription medicines during the children's last visit to Houston.

"Control your anger and emotions," he wrote. Then he turned to another matter. Piper had paid no child support since the court order had come down: "And please forward the $8,012.16 you owe me in child support."

Scraping up the money somewhere, Piper eventually sent the children airline tickets, at a cost of $879. When they visited, she and Jerry Walters took them on day trips to Galveston and Austin, where they swam in the city's favorite swimming hole, the chilly waters of Barton Springs. "We had a blast," he'd say later. "They were great kids, and Piper was absolutely engrossed with them."

Fred, however, continued to be angry. He called one day and Tina answered, telling him that Piper and the children were with Walters. "I'm unsure of the identity of 'Jerry,' and why I should expect the children to be with him," Fred e-mailed Piper, obviously peeved. "Is he another of your boyfriends or another of your lawyers?"

Yet, all too soon the children were on a plane, headed home to Virginia, and Piper was left in Houston, with more dark clouds gathering around her. It had been less than a year since she'd arrived in Houston, and her professional life was again falling apart.

Despite not collecting rent since a few months after she'd moved into the office, McVey had allowed Piper to stay. Then something odd happened: McVey began hearing that cases he referred to her, for which he was supposed to receive a percentage of her fees, had settled and she hadn't paid him. As if that weren't enough, one day a client called about a drunk-driving trial for the following morning. McVey quickly realized that Piper had taken a hefty deposit from the man, saying McVey would represent him, and then kept the money and told McVey nothing about the case.

"She had to go," he says.

Without explanation, McVey told Piper she had thirty days to move out of his offices. "She didn't even ask why," he says. "She knew what she'd done."

During her time there, Piper had put a sign in the front yard, one with both their names in raised letters. After she moved out, McVey blotted out her name with a thick coat of silver paint. Still, the legacy of her time with him continued, when a month or so after Piper moved out, a middle-aged couple came to McVey's looking for the lady lawyer who'd taken their money to represent them in a no-fault divorce. Wanting nothing more to do with Piper's problems, McVey gave the couple directions to her house, where she'd set up a small office, just blocks away. They left, and he thought little more of it, until Piper burst through his office door, furious. The couple, she said, had become infuriated when she admitted she'd never filed any of the paperwork. They were so angry, a shouting match erupted in the front yard, and someone had called the police to report a disturbance.

"Don't ever tell anyone where I live again," Piper ordered.

McVey agreed not to, but he was surprised. In his experience, attorneys usually wanted clients to find them.

About that time in Virginia, the judge held a hearing on Piper's back child support. She owed nearly $10,000, and the judge ordered her to begin paying an additional $200 each month to catch up. Judge Hammond also ordered her to either return Fred's mother's jewelry or pay him $4,000. When Bischoff's report on distributing the marital assets was released, the news, again, wasn't good for Piper. Rather than giving her half, Bischoff suggested the court award

Piper her own retirement funds, $7,858, and $67,487 from Fred's. It was just over twenty percent of the more than $300,000 in his 401(k)s. It must have been a particularly bitter blow to Piper, especially since Fred had offered her a more lucrative settlement a year before and she'd turned it down.

Still, her finances were looking up when, just before Christmas 2003, the court sent her the first check from Fred's 401(k): $32,212.29 from his account at the University of Richmond. Another check for more than $40,000 arrived just after the first of the year. With money to spend, Piper rented a one-story, white brick ranch house with a large backyard filled with trees in Kingwood, half an hour north of Houston. Her two cats had a deck to sun on, and inside she had a greenhouse and an aquarium filled with tree frogs. She decorated it with her antique writing desk and an armoire she bought along the side of the road. In the driveway, she parked her latest purchase, a new black Jeep Liberty.

Perhaps not surprisingly, the money went quickly, and by February, Fred was again threatening to take Piper back to court if she didn't pay him the child support due that month. In response, she sent nearly $2,500, which brought her up to date, along with a letter in which she claimed she was "unemployable even as a teacher" due to the divorce and not having custody of her own children. She then told him she expected him to use the child support money she'd sent to buy the children tickets to visit her over spring vacation. When Fred got the letter, he e-mailed her back: "It is amazing that you could send this to me as child support and then turn around and ask me to send it back to you to pay for visitation travel expenses!!!" He forwarded Piper's e-mail to Shilling with a note: "I can't quite believe Piper! She seems delusional and/or is playing games."

Yet, no matter what happened between them, Fred didn't stand in Piper's way when she wanted to see the children, as on Mother's Day when she flew into Richmond for the weekend and then wanted to extend the visit by a few days. Fred agreed, making up a list of the children's schedule for that weekend for her, from scouts to soccer. Despite the list, Fred later complained that the children never made it to their events. More than the other children, Callie seemed disturbed by being with her mother. This time when Piper briefly moved back into her life, Callie's Brownie leader and the school nurse both called Fred, saying his youngest was crying inconsolably.

That spring, Margaret Thatcher came to the Jepson School to lecture, and despite the continued battle with Piper, Fred's life and those of his children appeared to turn a corner.

Throughout the long divorce battle, Ciulla watched the way Fred had changed, becoming more thoughtful and in tune with his students. By that spring she was noticing that he finally appeared more at ease with the direction his life was taking. In the neighborhood, other parents noted how the Jablin children, too, appeared calmer and more at ease. "With Piper in Texas, it was like they were emerging from a long dark night," says the mother of one of Paxton's friends. "You'd see them in the yard with Fred, having fun and laughing. We thought that they might survive without too much damage."

Inside the house on Hearthglow, Fred covered the dining room table with boxes filled with family photos of Piper, himself, and all three of the children. On weekends, he sometimes sorted through them, organizing, as if trying to put in order the last twenty years of his life.

That March, 2004, Fred was invited to Rutgers University to give a speech. Instead of flying to the New Jersey campus the night before and being wined and dined as guest lecturers often do, he flew in the morning of the speech and returned the same evening, not wanting to leave the kids overnight. On the podium that day, he talked of his latest research. It would seem fitting to those who knew him. After all he'd been through the previous three years, worrying about what might happen to him and to the children, Fred had decided to investigate the concept of courage. Why? "Because it's an essential quality of an effective leader," he told the group gathered to listen. He talked of Germans who helped the Jews during World War II, and common citizens who changed their lives to help their communities. From the Latin, he explained, *courage* came from *cor,* the heart. One of the scholars he quoted was the philosopher Paul Tillich, from his book *The Courage to Be.*

"Grace strikes us when we are in great pain and restlessness," Tillich had written.

Meanwhile, in Houston, Piper's romance with Jerry Walters had flourished for more than a year. Off and on, there was even talk of love and marriage, but the passion had cooled by the spring of 2004, and they were dating others. Later, Walters would consider why. Tugging on the seam of his neatly pressed blue jeans, he'd recount how Piper loved children and animals, "anything that breathed." But at the same time, he felt she never considered his needs.

"Piper wanted everything to be about Piper," he says.

By then, like Tina, Piper had turned to the Internet to meet men, specifically the dating site Match.com, where she hooked up with Dean Lowry, who worked in the oil business. The first night they talked on the telephone, the call lasted four hours, but instead of becoming lovers, they became friends. They went dancing, something Piper loved, and Dean had a Harley-Davidson motorcycle. When they went out riding, Piper held onto him as they rode down country roads cut through the woods north of Houston. Once, when he came to pick her up, she was outside in her backyard, covered in mud, planting trees.

Dean listened to Piper's woes about the divorce, her conviction that she'd been misused and cheated of the children. Each day, she called her children, sometimes three and four times. To Dean, it was clear that more than anything Piper wanted to have them with her. When they came to visit, Piper had no time for him or anyone else.

Not long after they met, Lowry introduced Piper to Charles Tooke III, who worked as an oil landman, a freelancer who researched local, county, and federal records going back decades to document oil, mineral, surface, and royalty rights, helping oil companies determine who to pay for the crude they pumped. "I'd trained others over the years," says Tooke. "Dean asked me to show Piper what I did."

When Tooke received Piper's résumé on his e-mail, he replied, "You're overqualified. You'd be interested in this work for about fifteen minutes. It's way under you."

Piper called back and said she was just divorced and returning to Texas. "I'm in my mid-forties building a law practice, out of bullets and selling the gun for cat food," she said.

Tooke laughed, and made arrangements to show her the ropes.

They met at a restaurant on Lake Conroe, a dammed-up river turned into a pine-tree-lined lake an hour north of Houston. They sat outside at a table on the deck and talked. The son of an attorney, Tooke, who lived with his three dogs in an apartment near the lake, had graying hair and wore his glasses low on his nose. He had a soft Louisiana accent with a slight lisp, and a thoughtful manner.

That afternoon at the restaurant, he and Piper went over maps and records Tooke brought from a job he was working, and he explained how the work was done. There were interesting benefits. Piper could set her own hours and be paid by the day. If she worked full-time, she could earn $60,000 a year. The current project Tooke was working took him ninety miles south every day, to the Galveston courthouse, to do his research. Before long Piper was meeting him there, watching him and learning the ropes. Soon she was on her own, signed up as a contractor with the oil company on the Galveston project. When the checks started coming in she had the freedom that having money brought. The first time Tooke met Tina, she thanked him for helping Piper, saying that before he'd begun working with her, "Piper couldn't get her head up off the bed."

With Tooke, Piper rarely talked about Fred, and when Tina mentioned Piper's marriage, she told him that it had been to an abusive husband. A protective man, Tooke's heart went out to Piper. Before long he thought he sensed a spark between them, as men often did with her. When he asked her about going out together, however, Piper turned him down, saying she didn't date men she worked with.

"It's like church and state," she said.

Meanwhile, in Richmond, the children had become such a visible part of Fred's life that at UR the graduating class even gave him an award: "Most likely to leave the classroom during an exam to take a call from one of his children." Laughter filled the banquet hall when, as the award was announced, Fred was outside in the hallway talking on his cell phone with one of the kids.

Jerry Walters and Piper remained close, and that summer, 2004, he thought she was doing better than he'd ever seen her. She seemed to be thriving working with Charles Tooke, getting her life back together. She seemed happy, and she was painting again, including a portrait of the two of them together.

Finally, life seemed to be going well for both Piper and Fred.

In Richmond, Fred, too, was dating. Perhaps he'd heard of the Internet matchmaking craze from Piper, because somehow he ended up on the same website she and Tina had used, Match.com. Fred had dated one woman briefly, a school secretary, and a few women he'd met through a local dating service, but those relationships hadn't gone anywhere. That summer, he met a woman named Charlene, a single mom who lived in Ashland, not far from where Piper had rented a home just before her move to Houston. They'd started slowly, Fred telling people that he worried about getting involved. He knew he had a disappointing track record: two marriages and two divorces. Among worries that he had a tendency to pick the wrong women, he also considered the possibility that a woman might be attracted to him simply out of sympathy for a single dad with three kids. "I want someone who's really going to love me," he told a friend. "Not someone who feels sorry for me and wants to take care of the kids."

It was a happy time. One day, Fred and Charlene took their kids to Paramount's Kings Dominion, an amusement park with roller coasters and rides. By the fall when the children were back in school and Fred began teaching again, he and Charlene appeared to be getting serious, talking on the telephone each night before going to sleep.

In Harlingen, an old friend of Piper's heard that Fred was dating. She wondered what would happen. "Piper's children were Piper's children, and she wasn't going to share them with another woman," she says. "We all knew that."

Still, at least on the surface, all seemed to be on track for Piper that fall. She was caught up on her bills and was even current on her child support. She'd whittled down the arrears by nearly $3,000 to just over $7,000. Tooke enjoyed working with her, eating lunch in small restaurants in Galveston and walking through the Island City's many antique shops. At such times, Piper rarely talked about the children, although Tooke asked.

At the end of September, Piper e-mailed Fred and asked

if she could have the children during the Columbus Day weekend, and he agreed. Her plan was to fly to Virginia and take the children camping in the mountains.

Friday afternoon, October 8, the day Piper was scheduled to pick up the children, the security alarm went off at Fred's house, and Mel walked over to make sure everything was all right. She met Fred outside, and he told her he thought it was triggered by work at the next door neighbor's house. For a while they stood in the fall air and talked. He told her about Piper's plans to take the children camping in Washington State Park. "I don't know why she can't visit with them here and let them do what they have planned," he said. Paxton, twelve, was scheduled to participate in cotillion, a formal dance program he enjoyed, and that weekend, Callie, eight, had a soccer game. It was hardest on Jocelyn, fifteen, who had been asked to homecoming and now wouldn't be able to go. Their oldest wasn't happy about her mother's change of plans.

"I don't know why Piper can't do what other moms do, take Joce to get her hair and nails done, let her go to the dance and be happy for her," Mel commented.

Fred agreed. Of course they both knew that Piper wasn't like other mothers. When she was with the children, she expected their complete and undivided attention.

Still, Piper appeared to finally be maturing. She had a "real job," Fred said. And she was dating someone. More than during the past painful years, Fred appeared optimistic.

From there the conversation turned to Fred's new girl-friend, Charlene. He talked of their plans to attend the ballet that weekend. He smiled softly and looked happy. "I had the impression that it was getting serious," says Mel. "They were growing close."

Later that day, Piper arrived in Richmond, stopped at Steve Byrum's house—the friend who'd helped her pack for the trip to Texas—to pick up camping gear he'd agreed to lend her. From there she picked up Jocelyn, Paxton, and Callie at the Hearthglow house. Fred helped her load the car before they left to drive into the mountains, and later the children would say that their parents were friendly, if a little quiet.

Camping must have been invigorating that weekend. The Virginia scenery, always spectacular, was showing the first reds and golds of fall, and the air was crisp and clean. On Monday, Piper drove the children back down the mountain and returned to Steve Byrum's town house. By then Steve was on a business trip, and he'd asked Piper to put the gear away and leave his key in the house. Instead, she and the children stayed at Byrum's that Monday night. The next morning she drove the children to school, and then left to catch a plane home to Houston.

When Steve returned to his town house and found his gear piled on the pool table, he was furious. Reevaluating his friendship with Piper, he came to the conclusion that she was using him. "Every time she called me, she wanted something," he says. "I'd had enough of it."

Back in Houston after the trip, Piper and Charles were again in his SUV, traveling to Galveston to work. But something had changed, something about Piper. She was bubbly, excited about the fun of the weekend, the thrill of having spent time with her children. On the way home, Piper said she wanted to look for couches for her living room, and they stopped at the Foley's Furniture Warehouse, on I-45. Inside, they walked the aisles, laughing and mimicking an old married couple, Piper commenting on the couches and Tooke answering, "Yes, dear, you're always right."

"You'd make such good spouse material," she teased him.

"Well, you keep me in mind," he said.

That afternoon Piper found two beige leather couches and

a matching love seat for the living room. Charles lent her $1,000 to pay for them, and they loaded the couches in a rented trailer and drove them to her house in Kingwood. On the way there, Charles played a CD he'd just purchased: k. d. lang's *Hymns of the 49th Parallel*.

Looking back, it would seem to him that one song in particular struck a chord with Piper, Lang's mournful rendition of Leonard Cohen's "Hallelujah," with the verse:

> *Maybe there's a God above,*
> *But all I've ever learned from love*
> *Was how to shoot somebody who outdrew you.*

In Kingwood, Piper enlisted the aid of a neighbor to unload the couches. The furniture fit the room perfectly, wrapping around the fireplace. Afterward, Piper and Charles arranged and rearranged the room and sipped wine. Then Piper, who'd never before shown Charles a photo of her children, flashed dozens she'd taken on the camping trip on her computer screen. In the forest, the children were applecheeked and smiling, and as she and Charles looked at the photos, Piper shone with affection.

That Thanksgiving, Piper's sister Jean was marrying in Texas. And Piper and Charles planned what each of the children should wear. For Paxton, Piper thought a blue sport coat with a clip-on tie.

"No, get him a real one," Charles advised. "He needs to learn to tie one."

For Jocelyn, they decided, a little black dress, and for Callie something fun.

Not long after Piper and Charles sat down to eat carryout, soft music playing in the background, Jerry Walters walked in. Charles had heard about the on-again, off-again man in Piper's life, but he hadn't met him before. Although they'd dated others all summer, Jerry and Piper were getting close again. He wasn't sure where that would take them, but he enjoyed spending time with her. In Piper's living room, the two men had an uncomfortable moment, and Charles quickly left. After he was gone, Piper also showed Jerry photos from the trip, narrating the weekend for him. Jerry thought he'd never seen her happier.

Later, Charles would look back on that evening at Piper's and remember the stunning watercolors she'd painted of her children, scattered throughout the house. He was impressed with her artistic talents. But another of her paintings would come to mind: a painting of a solitary woman in a boat, rowing on an angry, bloodred sea.

In mid-October one of the neighborhood moms ran into Fred at the middle school, where Paxton was in the sixth grade. The previous May, Fred had planned the fifth grade graduation celebration for Paxton's class, with a party that included water balloons. It had all been great fun, and Fred made friends with many of the moms and dads. This day, he and the neighborhood mom started talking. Knowing Fred had been through so much turmoil within his own family, the woman easily confided in him, telling him she worried about her son, who'd been disruptive at home, refusing to do as he was told. Fred listened, then calmly said, "You know, it's going to be all right. He's a good kid, and he'll be fine." Something about the way Fred talked with her made the woman feel better, and she walked away thinking that after all Fred had been through, if he had faith in the future, shouldn't she?

It had been eleven years since he'd left Austin, but Fred was still in love with Halloween. Throughout the turmoil of the separation and divorce, the holiday had provided brief respites from the pain. The previous year, Fred had taken his children and Mel's oldest daughter, Chelsea, to Ashland Berry Farm for a hayride and to pick pumpkins. It was an all-you-can-carry pumpkin patch, and while the children laughed, slightly built Fred had managed to carry in his arms more than a dozen pumpkins, from small to large, when he crossed the finish line.

At home that October, 2004, as they did every year, Fred and the children got ready for the holiday, pitching in to rake the yard and stuffing faded jeans and an old shirt to make a scarecrow and pumpkin bags they positioned below the trees. There were carved jack-'o-lanterns and white-sheeted ghosts wafting in the crisp fall breeze. Bob and Doreen McArdle, the next door neighbors, watched, and were struck with how happy Fred and the children seemed.

They weren't the only ones that fall to see Fred with the children and believe that perhaps they'd come through the worst. Others noticed that Fred and the children appeared to be putting the pain of the divorce behind them. Even Mel didn't fret about her neighbor and friend as often. And she'd stopped worrying about Piper and Tina, fearing that they might do something to Fred or the children. "It all seemed to have worked out," she says. "Fred was hopeful, we were all hopeful, that the worst was over."

Meanwhile in Houston, that Monday night before Halloween, Jerry picked up dinner at La Madelaine, a small French, wood-paneled restaurant, and brought it to Piper's house in Kingwood. They ate and went to bed. The next morning, Tuesday, Jerry was struck again by how happy Piper seemed, as if she didn't have a care in the world. He kissed her goodbye and left.

The following morning, Piper left for work in Galveston. She brought with her Tina's live-in boyfriend, Mac. The DJ was taking a break from the broadcast business and considering a new career. Piper agreed to teach him the oil business, training him as a landman, as Charles Tooke had trained her. Monday and Tuesday, they traveled back and forth together to Galveston, calling Charles often to ask his opinion. Mac would later say he was disappointed when Piper told him they wouldn't be able to work that Thursday and Friday. Yet he understood when she explained that she had a legal continuing education seminar to attend, in Fort Worth.

Everything seemed so calm that last week in October, 2004, so normal. Nothing seemed particularly out of the ordinary. But looking back, there was something, perhaps, that might have hinted at what was to come: Later, Mac told Charles that the Rountree sisters irritated him, listening over and over to one song on a CD Charles had given Piper, the same CD he'd played for her the day they bought the couches—k. d. lang's *Hymns of the 49th Parallel*. The song that Piper and Tina played over and over was the same one Piper gravitated toward the first time she heard the CD: Leonard Cohen's "Hallelujah," with the verse:

> *Maybe there's a God above,*
> *But all I've ever learned from love*
> *Was how to shoot somebody who outdrew you.*
> *It's not a cry you hear at night*
> *It's not somebody who's seen the light.*
> *It's a cold and it's a broken hallelujah.*

10

....

Darkness cloaked Hearthglow Lane on the morning of Saturday, October 30, 2004. At just after six two golden retrievers began a troubled, angry barking across the street and two houses down from the Jablin household. The dogs' owner, groggy after being awakened from a deep sleep, wondered if a raccoon could be searching for garbage cans to topple, a breakfast of discards to claim. Bundled in her bathrobe, the heavyset woman walked the dogs to the back door and out onto the back deck. Once outside, as inexplicably as they began barking, the dogs stopped. Quiet restored, the woman retraced her steps, ordering the dogs back inside the house. With that, the woman locked the back door and returned to her bed.

Half an hour later, at 6:37 A.M., Bob and Doreen McArdle lay in bed awake on the second floor of their house, directly next door to the Jablin house, the windows open to air out the newly painted bedroom. Then, suddenly, three shots rang out through the streets of Kingsley, piercing the early morning silence. Bob, a gray-haired salesman for a consulting company, jumped from bed and ran to the open window. The bedroom overlooked the street, and he saw a shadowy figure—he couldn't tell if it was a man or a woman—running left to right across his front lawn, from the same direction as the shots.

Wondering if the sound could have been from construction

workers remodeling a house down the street or a teenager firing a gun in the woods, Bob fumbled for his glasses. No, he thought. McArdle was a former Marine, and to him it definitely sounded like gunshots, and it sounded closer than the woods.

Grabbing the phone, he dialed 911. "Someone just shot a gun off, and it sounded like it was close," said Bob, a transplant from New York, in his Long Island accent.

"We have a squad car on its way," the dispatcher said.

With that, Bob and Doreen, a petite woman with short tousled brown hair, roused from bed, grabbed robes and walked downstairs to wait for police. Within minutes they saw a squad car circulating through the street, pausing in front of their house and others, shining flashlights onto the darkness, illuminating lawns, driveways, and front doors of houses as they crept through the neighborhood. Bob waited, watching, until, ten minutes after his call, the squad car pulled up in front of his house.

"Find anything?" Bob asked when he opened the door.

"No." Henrico County police officer Philip Maggi, a square-jawed man dressed in a blue-shirted uniform, shrugged. "Tell me what happened."

Again, as he had to the dispatcher, Bob McArdle told of being in bed when he heard three gunshots. "It sounded like it came from somewhere down there," he said, pointing to his left, down the street, in the direction of the Jablin house.

"Okay," Maggi said. With that, he left, and the McArdles went back inside, and again watched the officers search. In the pitch-darkness, the police flashlights skimmed the street, the grass, the houses, throwing slim slivers of light. The day before Halloween the yards on Hearthglow were decorated for trick or treat. When the light from the flashlights illuminated leaf-bag pumpkins and wind-tossed scarecrows, it gave the street the look of an eerie tableau.

It was still black outside when, at 7:00 A.M., half an hour after he heard the shots, Bob answered the door a second time. "We didn't find anything," Officer Maggi said. That wasn't unusual. Reports of sporadic shots fired were relatively common, and often the source was never found.

"Okay," Bob said. "Don't worry about it. My wife and I are going for a walk with the dog. We'll take a look around once it gets light. If we find anything, we'll call you back."

With that, the morning's excitement appeared over. Maggi and his partner drove off in the squad car, and Bob and Doreen McArdle returned to their second-floor bedroom to dress for their morning dog-walking ritual.

At seven-fifteen, as the darkness lifted, the McArdles exited the house out their back door, with the dog on a leash, and walked out toward the street. It had rained the night before, and the wet street and grass glistened. The breeze was crisp and smelled of fall. Just as they walked in front of the Jablin house, another neighbor drove by and Bob waved to stop him. "Did you hear anything this morning?" he asked.

"Yeah," the man said.

"Where did it come from?" Bob asked.

"Sounded like around here," the man said.

It was then that Doreen glanced toward the Jablin house. As the darkness faded and the sun crept upward over the horizon, she looked up the driveway's incline, toward the level area near the house, and saw something crumpled on the driveway, below the family basketball hoop and next to Fred's black 1999 Ford Explorer.

"What's that?" she asked.

"That's not anything," the man said, before he drove off to the golf course, appearing worried about missing his tee time.

The two of them alone again, Bob wondered if what he saw could be something as innocent as a Halloween decoration, perhaps a scarecrow that had blown out of a front yard and onto the driveway.

"You'd better check and see what that is," Doreen said, with a feeling of dread.

Holding the leash, his dog sauntering beside him, Bob walked up toward the garage and Fred's Explorer, along the right side of the Jablin driveway. As he drew closer, he saw a figure with a bleached white face, paper-white ankles protruding from navy blue sweatpants, and too-white hands and wrists. The figure appeared starkly pale, as if bloodless. What had been obscured by the darkness was now easily seen.

That's no scarecrow, he thought.

Not wanting to, wishing he could turn around and leave, Bob cautiously walked closer. There was no mistaking what he'd found, who was lying there. His heart pounding, he turned back toward Doreen, who stood on the street holding her cell phone. "Call the cops back," he shouted. "It's Fred."

At 7:25 A.M., less than an hour after they awoke to the crack of gunshots, Doreen dialed 911. This time she left no room for conjecture. "We found Fred Jablin, our neighbor, in his driveway," she told the operator. "It looks like he's dead."

Within minutes Maggi and his partner roared back up Hearthglow Lane in their squad car. They ran from the car, and Doreen pointed toward where Bob stood. Maggi sprinted up the driveway and crouched beside the body, feeling for a pulse, finding none. Bob McArdle glanced around and noticed Fred Jablin's eyeglasses, lying on the ground, as if they'd flown off as he fell.

Quickly the call went out: Police and an ambulance were needed at 1515 Hearthglow Lane. From that point on it seemed to the McArdles that time both stood still and sped up, as what they'd anticipated to be a relaxed Saturday morning took on an unnatural urgency. Squad cars and an ambulance flooded the street, and officers strung crime scene tape across the front of the Jablin house. Bob McArdle saw an officer in his yard pointing at footprints in the wet grass, leading from the body in the driveway past his house.

"That's where I saw someone running," Bob told the officers. "Right after I heard the gunshots."

By then paramedics had lifted Fred's lifeless body onto a thick blue plastic backboard. They cut his blue sweatshirt up the front and peeled it back from his chest. When they rolled him over, they found a bloody hole in the back of his sweatshirt that lined up with a bullet wound in his lower back. His body was cool and still, yet they checked for a pulse. They found none. As the paramedics worked, police congregated outside the house, looking through the yard, searching the neighborhood. One paramedic injected heart-stimulating drugs into Fred, and another breathed into his lungs and pushed on his chest, administering CPR. It was too late.

Fred Jablin was dead.

Once on the scene, Sergeant George Russell began pointing to officers and issuing orders. A black-suited SWAT team arrived after the McArdles and other neighbors warned that three unaccounted for children lived in the Jablin house. Were they being held hostage? Was the gunman still close, watching from the trees, or inside the house with the children? No one knew.

Officer Robbie Reamer, a nineteen-year veteran with the force, who'd arrived on the scene early and was assigned to cordon off the street, was called back to the house and told he'd be going inside with the SWAT team, to find and if necessary rescue the Jablin children.

Later, Officer Reamer would remember entering the house that morning as if in snapshots, walking through the unlocked back door and into the kitchen, where he smelled the rich aroma of the coffee Fred had set to brew that morning still steaming in the pot. Carefully, the officers cleared each room, walking past the dining room table covered with the family photos Fred was organizing, and finally up the stairs to the bedrooms.

The first bedroom the officers approached was Jocelyn's, but the door was locked. Seconds later they heard the door being unlocked, and the pretty teenager with her soft brown hair in her sleepy eyes emerged from her bedroom.

"We have to get you out of the house, come with me to the back door," Reamer told her. He walked her down the steps to the door.

"What's going on?" she asked.

"I can't tell you now," he answered. "You just need to get out of the house."

At the back door, Reamer handed Jocelyn off to Maggi. "Stay with this officer."

"What's going on?" she asked Maggi.

"We'll explain later," Maggi said.

"Make sure my brother, sister, and dad get out," she said as Reamer walked away, grateful that the Explorer blocked the view of the body.

Back upstairs, the second team had moved into Callie's room to wake her. They nudged the youngster several times. She didn't respond. For a moment Reamer feared the eight-year-old was hurt or dead, but then, gradually, her eyes opened. Again the question no one wanted to answer: "What's going on?"

Reamer was the first into Paxton's room, and shook him. The twelve-year-old's eyes opened wide, and he looked instantly frightened. "Come on, buddy," Reamer said. "We have to get you out of the house."

Paxton shook off sleep and saw Reamer in his police uniform and the other uniformed officers behind him.

"Okay," he said.

With that, Reamer and the other officers escorted the two youngest children downstairs. Once there, all three children were hustled out onto the deck their father had built years earlier. Then they were directed to turn left, away from the driveway, the family's black Explorer, and their dead father's body. Officers lifted them over the split rail fence on the side of the house that led to the McArdles' front yard, and rushed them toward a rescue vehicle waiting in the street.

When Henrico County investigator Coby Kelley arrived on the scene, Fred Jablin's body still lay on the backboard, covered by a white sheet, and Jocelyn, Paxton, and Callie had been put in an ambulance and driven down the street, out of eyeshot of the house and the driveway.

In Homicide for nearly four years, Kelley was a broad-chested, six-foot-two-inch, thirty-two-year-old man with a short, military-style haircut. He'd grown up ten miles from the Jablin house and had become a cop for many reasons, not the least of which was that he enjoyed "sitting across from somebody and getting them to say things that would probably put them away for the rest of their lives." An investigation, in his eyes, was a game, one with incredibly high stakes. Despite his case-hardened attitude, Kelley, who'd once wanted to be a lawyer, had a boyish look and an infectious enthusiasm for "the job," as police refer to their calling.

In the street, Kelley met with the first officers on the scene, Sergeant Russell, and their boss, Captain Jan Stem. The minute Stem, a dapper man with glasses and reddish-brown hair swept over his forehead, heard the deceased was a University of Richmond professor, he knew the pressure would be on. "The murder of anyone that prominent in the community is bound to cause a stir," he says. After thirty years with the Henrico P.D., twenty-six in investigations, he figured he'd pretty much seen it all.

Henrico County had changed over Stem's tenure on the

force. In the beginning it was a sleepy place to work, but as Richmond's sprawl reached north and the county grew, he'd become busier and busier. In the past, third-shift officers rarely encountered more than teenage pranks. As he'd climbed the ranks, the number of officers grew from two hundred to more than five hundred, and the night shift tackled many of the same crimes as those within the city limits, including drug cases and murder. Still, not usually on the quiet, upper-class West End.

As the uniformed officers filled Kelley and the captain in, Stem's head started spinning with the possibilities. Maggi and others had already heard about Fred Jablin's nasty divorce, and, of course, an ex-spouse was always an intriguing murder suspect, but there were so many others. Could it have been a burglary? Stem didn't think so, when he heard there was no sign of a break-in and that, at least on first inspection, it didn't appear the house had been ransacked or that anything was missing. What about Fred's students at UR? Could one of them have been angry enough to want their professor dead?

Stem knew that the first forty-eight hours in a murder investigation were the most crucial. The Henrico P.D. had a policy they'd initiated in 1997, one that served them so well they'd cleared all but eight of the seventy-three homicides in the county since 2001. For the first two days, they flooded a murder case with staff and resources, everything available, to make sure no leads were left unexplored. "You've got to hit a homicide case hard, with everything you've got," says Stem. "It's the only way, hard and fast, because we want to lock up the bad guys."

Before he proceeded any further on this particular case, however, he had a problem: what to do with the Jablin children, who remained in the rescue unit, parked on the street? Family would be contacted, but that would take time. When the subject came up, Officer Harry Boyd cut in, explaining that he and his wife, Barbara, lived two blocks away from the Jablins, and that they knew the Jablin children. Callie and his son were on the same soccer team and went to school together.

"I'd be happy to bring the kids to our house, to stay with my family, while we sort this out," he said.

Stem thought for only a moment before agreeing. The children were going through enough, he figured. Why not at least let them stay with someone they knew? But Boyd wouldn't be the only officer there to talk to them. Stem picked up the radio and called into headquarters. "Send the special victims unit out here to talk to the Jablin kids," he ordered. "We need to know what they can tell us."

That settled, Stem felt certain of two more priorities: He needed access to the inside of Fred Jablin's house, pronto, and he wanted to make sure it was done properly, so that whatever he found inside would one day be admissible in a court of law. "Hanna, get us a search warrant for the house," Stem ordered Investigator Chuck Hanna, an African-American officer with a broad smile and an athletic build. As soon as Stem barked the order, Hanna took off, understanding the urgency and eager to comply.

The morning was moving quickly; by eight-thirty the Jablin children were at the Boyd house, and Stem had Hearthglow Lane crawling with uniformed and plainclothes officers. Some searched sewers, looking for the murder weapon. In the driveway, crime scene specialist Danny Jamison began collecting samples from around the body. Jamison, a gray-haired officer with arched, dark eyebrows, had been on his way to help build a Habitat for Humanity home with his wife and daughter when the call came in that he had a murder scene to process. He'd arrived quickly and gotten right to work, swabbing blood on the driveway, hoping it might have the DNA of the killer. In plastic bags, he placed Fred's eyeglasses and a "to do" list another officer discovered on the driveway near the Explorer. It had a list of handwritten instructions, including taking the car to the shop and picking up the laundry.

Throughout the morning, Captain Stem and Sergeant Russell acted as coordinators, deploying the manpower on the scene. On their orders, officers fanned out to canvass Hearthglow, talking to neighbors. Stem had already contacted Wade Kizer, the commonwealth's attorney in charge of the county's staff of criminal prosecutors. Stem reached Kizer while he was out to breakfast with friends. "We've got a man shot dead in his driveway on the West End," Stem told him.

"I'll be right there," Kizer said.

While the others worked Hearthglow and Hanna hurried back to headquarters to type up the request for the search warrant, Kelley decided to talk to the people who'd found the body. The McArdles stood nearby, still stunned by their early morning discovery, not wanting to be there but unable to tear themselves away from the horror unfolding around them. Quickly, they explained to Kelley what had happened that morning, then turned the conversation to Fred's divorce. "You need to talk to Melody Foster. She lives directly behind Fred," Doreen suggested. "She was close to Fred, and she can tell you a lot more about Piper than we can."

"It was a horrible divorce," Mel told Kelley, after he'd rung her doorbell and introduced himself. Just a short time earlier, Pete had come into the bedroom to tell her what had happened. Mel and Pete had both slept through the shooting, and he'd learned of it from a neighbor, who'd seen the children being shepherded by SWAT officers out the back door. What would happen to Jocelyn, Paxton, and Callie now? Mel wondered, shivering from the news. Wanting to help, she gave Kelley everything he asked for, including the name of the woman Fred was dating, Charlene; Piper's ex-paramour, Dr. Gable; the Jablin children's nannies; and the information on how to find Piper's best friend in Richmond, Loni. Throughout the conversation, Mel explained to Kelley the turbulence of the Rountree-Jablin divorce, including telling him of Piper's false domestic violence charges against Fred. Piper, as Kelley was learning, was capable of nearly anything to get what she wanted.

"For a long time, Fred was afraid of Piper," Melody said. "But lately, things seemed better. He was hoping maybe the worst was over."

By the time Wade Kizer—who'd worked at the commonwealth attorney's office for twenty-five years and had led it for four—arrived on the scene, the coroner's office was getting ready to remove Fred Jablin's body. Pictures had been taken, and Jamison, the lead crime scene investigator on the case, had inspected and documented the area. He'd found one copper-jacketed bullet on the right side of the body, in the grass, but no shell casings. After checking in with Stem, Kizer—a balding man with a monk's fringe of brown hair, glasses, and a halting manner of speech—talked to Sergeant Russell. "We need to send someone to the university to secure the victim's office," said Kizer. "We don't want anyone going in and out, nothing disturbed, until we've had a thorough look." Russell nodded, and dispatched two officers to the University of Richmond.

Word had already reached the campus that morning that someone had been killed on the West End, and that the victim could be Fred Jablin. A UR employee who lived near Kingsley had even driven by and seen what appeared to be Fred's body in the driveway. Calls went out to faculty, and a group gathered to contact Fred Jablin's students. Dean Ruscio didn't want students hearing on the radio or television that their professor was dead. When Ruscio called Joanne Ciulla to tell her Fred had been murdered, she blurted out, "Oh my God. Piper killed Fred."

Two years earlier, a group of officers had come to the Jepson School to arrest Fred Jablin. Now, late on the morning after his death, a second group arrived, asking to be directed to his office. Once there, they installed their own lock and strung yellow crime scene tape across the door.

At 10:00 A.M., Captain Stem's phone rang. Hanna was on his way back to the crime scene with the signed search warrant in hand.

"We're in," Stem told Russell. "Let's get going."

With that, Kelley, Jamison, and a few select officers moved into the two-story house in which just four hours earlier Fred was brewing his morning coffee and getting ready to take the children to the neighborhood pumpkin festival. Later, neighbors would remember how the year before Fred had asked them to take pictures of him with the children. "In the past, I always took the pictures of Piper with the kids," he'd said, handing them his camera. "We don't have many of me with them." Now, those few photos would become precious, holding dear, never-repeated memories.

In the driveway, the coroner's technician was inserting Fred Jablin's lifeless body into a black vinyl bag. Moments later the corpse was loaded into the back of the M.E.'s hearse, to be transported to the morgue for autopsy. Jewish custom disdains disturbing remains, but even if someone in Fred's family had complained, the result would have been the same. His body was a prime piece of evidence, and until they were finished with it, within the jurisdiction of the Henrico County Police Department.

At the Boyds' house, the Jablin children were being gently questioned by Investigator Judy Berger, of the Henrico P.D. Special Victims Unit. It had taken time. Before officers could begin, their rabbi, Martin Beifield, had been called to come to the house from Congregation Beth Ahabah. It was a sad duty. Just the night before, Fred had been at the synagogue with the children for Friday night services. When he told them of their father's death, the three children sobbed, at first uncontrollably. Finally, after much coaxing, they settled down, and Investigator Berger was able to talk to them. Afraid to again upset them, she proceeded slowly.

Of the three children, Jocelyn was the only one who'd heard the shots. Not knowing what they were, she'd rolled back over and gone back to sleep, never dreaming that her father could be dying just steps from their own back door. When Berger asked if she and Fred had fought, or if she had a boyfriend he didn't care for, someone he was trying to push out of her life, the fifteen-year-old answered unequivocally.

"No," Jocelyn said. "My dad wasn't mad at me, and I don't have any boyfriends."

All the children insisted their relationships with their father were great. There were no problems.

Nothing out of the ordinary had happened the night before, they all agreed. Fred had made them dinner and then taken them to the synagogue. They returned home about ten, and immediately got dressed for bed. When they were ready, Fred tucked each of the children in and kissed them good-night, as he did every night. Saturday was to be a busy day. Callie and Paxton each had soccer games, and then there was the neighborhood pumpkin festival at Gayton Crossing Shopping Center.

Such ordinary plans, and then something extraordinarily tragic had changed everything.

It would prove to be an emotional and difficult morning at the Boyd house, as Berger led the children through the questions she needed answered. The children knew of no one who was angry with their father. At first they weren't sure when their mother had called them last. "She calls a lot," they said. Then Paxton remembered Friday afternoon, after school, when he'd been in a friend's garage with a group of his friends playing poker, like the celebrities on television. His mom had called, and he remembered talking to her, and then she talked to Russell Bootwright, one of his best friends.

"Where was she when she called?" Berger asked.

"She was in Texas, driving home from working in Galveston."

"Did she sound mad at your father?"

"No," Paxton said.

"What did she talk about?"

"A raccoon living under her porch," the youngster said. When Piper found out the boys were playing poker, she laughed and said she'd call later.

Callie, too, had talked with her mother on that Friday, at about six-thirty that evening, when Piper called the house. Was she sure it was her mother? Yes, the little girl said, she was sure.

When Berger asked what their parents' relationship was like, she was told "they get along." The last time they'd seen their mother, they said, was a couple of weeks earlier, when they'd gone camping. They'd borrowed the camping gear from one of their mother's friends, a man named Steve Byrum. Steve wasn't a boyfriend, just a friend, but their father did have a woman he was dating, they said, a woman named Charlene.

Inside the Jablin house, Jamison noted that the house alarm had been turned off. He found the keys to Fred's Explorer and two cell phones resting in cradles in the kitchen, charging. Picking through the house, Jamison looked for anything that appeared out of place, anything that could be remotely important to the investigation. One officer found a photo in the children's room of a woman in a red dress, a woman neighbors would identify as Piper Rountree, Fred Jablin's ex-wife. The officer confiscated the photo for later use. Meanwhile, in the kitchen, Chuck Hanna, who'd arrived with the warrant, flicked through the recent calls on the home phone and the two cell phones. On both phones he found a number that ended in 7878, one that displayed as "Mom's Cell." Inside the pantry, he found a phone number list, and there it was again: "Mom's Cell" with the 7878 number.

At about the same time, Investigator Berger called Captain Stem from the Boyd house. "The kids are doing well," she told Stem. "They say they talked to their mother yesterday and that she was calling from Texas."

"What kind of phone was she on?" Stem asked.

"I'm not sure," Berger said.

"Find out," he ordered.

"Look at this," said Hanna, who'd wanted to be a police officer since childhood, when he met a friend's detective father. Hanna showed Stem the cell phone with Piper's phone number displayed on the screen. "Looks like the kids' mom called yesterday on her cell phone."

Just then, Stem's own cell phone rang again.

"She called the kids on her cell phone," Berger confirmed.

"Bingo," Stem said with a smile. "Get Cindy Williamson working on this. We need to subpoena the records for the ex-wife's cell."

Williamson, a middle-aged woman with shoulder-length, graying brown hair and oval glasses, worked in the commonwealth attorney's office as an expert at procuring and dissecting records, including bank and phone records. Already in the Henrico P.D. office organizing records for another case that Saturday morning, when Stem put in the call, asking her to shift over to the Jablin investigation, she didn't hesitate.

On the scene, Kizer offered to head back to headquarters. "I'll help Cindy write up the request for the subpoena," he said. It wasn't his favorite way to spend a Saturday. Henrico's head prosecutor would have much preferred to have been with his wife and two sons, or saltwater fishing, feeling the bite of the sharp ocean air on his face. But that wasn't an option; a man named Fred Jablin had been murdered, and Kizer wanted as quickly as possible to find the person responsible. Perhaps Jablin's ex-wife had nothing to do with his murder, but if she did, tracing her phone calls could turn out to be an important piece of evidence.

At about that same time in Houston, the phone rang at Doug McCann's town house, in the Memorial area, just west of the city. "Do you know who I am?" a woman asked.

"No," McCann admitted.

"Well, it's Piper. I'm looking for my sister."

This had been an unusual weekend for McCann, the head of a Houston oil and gas manufacturing company. He and Tina had met about a year earlier on Match.com, dated briefly but then broken it off. Yet just the day before, Tina had called McCann, suggesting they get together for dinner. He'd agreed, and Tina ended up staying overnight at his town house. Doug knew about Mac, but Tina had told them they were just friends.

"Tina just left here," he told Piper. "She's on her way to work."

"I was trying to see if you two want to go out to dinner tonight," Piper said.

"Well, sure, if we can work it out," Doug said.

"How did you like that red dress Tina wore last night?" Piper asked. "Some say it looks better on me."

Doug didn't answer.

"Do you think I sound like Tina?" she asked.

"Maybe," he said.

"Some people say we sound alike," she said before she hung up.

At eleven o'clock that morning, Stem's phone rang on the scene.

"I've got the first results back on Piper Rountree's cell phone from Sprint," Williamson said. "They can't say where she is today yet, they're working on that, but yesterday someone made calls on that cell phone that bounced off towers in Richmond."

"Well, isn't that interesting," said Stem. "Keep me posted."

"Will do," Williamson said.

When he hung up the telephone with Williamson, Stem thought about what that piece of information meant. Rountree's phone was in Virginia, but that didn't necessarily mean that she was. He decided it was time to make a second call, and minutes later he was talking to Houston Police Department Lieutenant Rick Maxey, in charge of the homicide department that Saturday. Stem explained the situation, that Rountree was—as the ex-spouse always is—someone they were interested in, and he asked Maxey to try to locate her. "Assuming she doesn't know, we need to fill her in and tell her we have her kids," said Stem. "We'd, of course, like to verify if she is in Houston, so we can rule her out as a suspect."

In Houston, Investigator Breck McDaniel and his partner, Sergeant David Ferguson, were the next up for a case, and Maxey responded by immediately assigning them to help out the Henrico P.D. McDaniel, a husky, jowly investigator with short brown hair, and Ferguson, in his mid-fifties and a twenty-nine-year veteran of the force, pulled up Piper's information on the computer. They found a driver's license with her photo, blew it up, and printed it off in black-and-white, along with information on the state car registration for her black Jeep Liberty and her address in Kingwood. After getting her phone number from information, they left the Houston P.D.'s downtown headquarters and drove out on I-59, through the maze of skyscrapers, north toward Kingwood. Once they arrived, they rang the bell and knocked. No one answered. They called the telephone and heard it ringing inside, but no one picked it up. McDaniel looked in the garage window and found it empty.

At 12:49, before they left Piper's house, the investigators did one more thing: McDaniel called Piper's cell phone. No one answered, but he left a message, asking her to call Coby Kelley with the Henrico P.D.

Back in Virginia, Loni had heard about Fred's murder while watching her children play soccer that morning. When she showed up at the Jablin house, she was screaming and nearly hysterical. Chuck Hanna pulled her to the side and coaxed her back into her car, telling her the children were safe and there wasn't anything she could do to help. "Was Ms. Rountree in Virginia this weekend?" he asked.

"No, I don't believe so," Loni said. "I haven't talked with her."

"She didn't stay with you this weekend?" he prodded.

"No," Loni replied. "I haven't spoken with Piper in a month or more, and I don't think she was even in town."

"If you hear from Piper Rountree, tell her we need to talk to her," Hanna said, before Loni drove away.

By then, Cindy Williamson was again on the telephone with Captain Stem. "We've got more information from Sprint," she said. "The phone was used here this morning, just after 4:30 A.M., and it bounced off a tower on Cox and Broad Streets, about five miles from the Jablin house."

"Looks like Rountree was here," Stem told his officers. "We need to find out where she stayed last night and how she's traveling, if she had a gun. We've got a lot of questions to answer."

At two-thirty that afternoon, Henrico police had tracked down a number for Michael Jablin in northern Virginia. When they put a call through and told them what had happened, he was at first quiet, apparently stunned. Then he said, "Have you considered my brother's ex-wife as a suspect?"

When he hung up the telephone, Michael couldn't believe his brother was dead. Hoping it was all a bizarre mistake, he dialed Fred's cell phone number and listened to it ring. But no one answered.

By then Chuck Hanna was ringing the doorbell at Steve Byrum's town house, in the Sussex Woods section of Richmond, just down Gayton Road from Kingsley. Byrum was upstairs watching football on television with his son when the doorbell rang, and came downstairs reluctantly when he saw men in suits, wondering if there were missionaries at the door wanting to convert him. When one of the officers at the door flashed his badge and asked about Piper Rountree, Byrum didn't look happy. "She's been calling me the last couple of days, but I haven't been answering," Byrum said. "I'm pretty upset with her over the last time she was here. Why are you looking for her? What's happened?"

"Have you seen her this weekend?" Hanna asked. "Did she stay with you?"

"No," Byrum insisted. "Tell me what's happened."

Hanna left without explaining.

At the mall, where her son was in a karate exhibit, Fred's girlfriend, Charlene, talked with yet another investigator. The news had shaken her badly. She and Fred had begun e-mailing in June, after meeting on the Internet website Match.com, she said. The last time she saw him was for lunch the previous Wednesday, and they'd talked on the telephone a little after ten just the evening before, as they often did before going to bed. She told the officer that there'd been some back and forth between Piper and Fred recently about whether he was willing to let the children go to Texas to attend Piper's sister Jean's wedding over Thanksgiving. "He decided he'd let them go," Charlene said. "But he'd been e-mailing Piper about money again lately, looking for the child support she owed him."

Meanwhile, back at the Jablin house on Hearthglow, now known as the murder scene, Captain Stem was getting yet another call from Cindy Williamson. More Sprint records were being faxed in. These were even more enticing. Calls had been made off Piper's cell phone early that morning that showed it had traveled southeast on Interstate 64, leaving Henrico and cutting through the heart of Richmond. At one point it crossed the James River, a favorite place for local criminals to dispose of weapons and evidence. But there was more. "We know that around noon the phone was used in Norfolk," Williamson said. "Then—this is interesting— an hour and a half later, it was in Baltimore."

"Oh, yeah. We've got a good lead now," Stem crowed.

Kelley, Stem, Russell, Hanna, and the others discussed the possibilities. It was apparent that the person on the telephone hadn't driven the 240 miles from Norfolk to Baltimore in such a short time. "She's traveling by air," Stem said. Getting on the telephone, he called headquarters and asked for an investigator to drive to the Norfolk airport to try to find out if anyone there remembered seeing Piper. With Homeland Security in force after 9/11, Stem figured she must be traveling with her own identification. "We need to track down a ticket for a Piper Rountree or a Piper Jablin."

Minutes later Williamson called again. "I just got a fax from Houston," she said. "They're saying there's not just a Piper but a Tina Rountree. Maybe your officer should check the airline records for that name, too."

* * *

At three that afternoon Chuck Hanna and another officer drove to Loni's house, built on a ravine just five minutes from Hearthglow. Loni had married, and she and her husband, Bill Gosnell, an attorney, were renting the frame house while they built a new one. Again Hanna asked Loni if she'd heard from Piper. Had her friend stayed with her that weekend? "No," Loni answered to both questions. Hanna left unconvinced. Soon after he did, Loni picked up the telephone and punched in Piper's cell phone number. No one answered. Throughout that day she dialed the same number over and over again. Each time, she was diverted to voice mail. Piper wasn't answering.

"Call me, Peeps," Loni said. "It's important."

Around that same time, Kelley was also calling Piper's cell phone and leaving a message. His was more forceful: "Ms. Rountree, this is Investigator Coby Kelley with the Henrico Police Department. We need to talk to you. Please call me at . . ."

At the University of Richmond, many of the calls had already been made to Fred Jablin's students, to tell them of their professor's death. "We thought at first maybe it was a heart attack," says one student. "But then we knew about the divorce, and a lot of us wondered how he'd died. No one would tell us." A meeting was scheduled at the school for Sunday morning, with the faculty and chaplain. More information, they were told, would be forthcoming.

When Captain Stem's phone rang about 3:30 P.M., it was the officer he'd dispatched to the airport. "I didn't find a record for a Piper Rountree, but a Tina Rountree flew out of Norfolk on Southwest Airlines flight 247 at 12:30 P.M., changed planes in Baltimore, and is right now on Southwest Flight 2609 that left at two-ten, on her way to Houston, and scheduled to arrive at four-thirty central time," the officer told her.

First Stem called Kizer to tell him of the break in the case. Someone, a woman, was traveling back to Houston. Stem wanted to have the Houston P.D. get a warrant for her arrest and to hold her, to confiscate her bags and search them for evidence. "Wade, we know she was here in Virginia at the time of the murder," he said. "And if she has evidence on her, she could destroy it."

As much as Stem argued, Kizer wouldn't budge. "All we have at this point are some cell phone records. We don't have enough to do that," he said. "Let's just send someone to the airport to see if they can verify who is getting off that plane, Tina Rountree or Piper Rountree, then seize her bags and hold them until we can get a warrant." Kizer and Stem went way back together, to a time when they'd both worked Burglary, and they'd had this type of disagreement often. As a police officer, Stem often wanted the suspect taken in and charges processed before Kizer felt he had enough for an indictment. He knew he'd have to try a case in a courtroom before a judge, where *how* evidence was obtained could become a paramount issue.

Reluctantly, Stem agreed. As soon as he hung up with Kizer, he called Houston and asked to have officers sent to the airport with photos of Piper and Tina Rountree, to identify the sister traveling as she got off the airplane, seize her luggage, and hold it until they had a warrant.

At Henrico P.D. headquarters, Cindy Williamson was following yet another paper trail: the information from Southwest Airlines that showed Tina Rountree had charged the airline ticket on a Wells Fargo debit card. Once she had that detail, she picked up the telephone and called the bank's credit card division. The supervisor on duty, however, refused to cooperate without a subpoena signed by a judge.

When Williamson told Kizer, he stopped what he was doing to draw up the paperwork.

Meanwhile in Kingsley, Ana, the Jablin children's nanny throughout most of the tumultuous divorce, left the Boyds' house with Investigator Berger, after comforting the children, to go to the house on Hearthglow to claim enough of their clothing to get them through the coming days. On the scene, the nanny told Kelley about her run-ins with Tina Rountree, how explosive the woman could be. Crying and upset over Fred's murder, she choked out the words as she described Piper as unstable.

In Houston, Breck McDaniel and David Ferguson, six-foot, his brown hair balding, with a soft Louisiana drawl, had been getting off work for the day when the call came in that the Henrico police needed someone to meet a Southwest Airlines flight from Baltimore at Houston's Hobby Airport, the airline's hub, just south of downtown. Armed with grainy driver's license photos of Piper and Tina, they left for the airport. McDaniel, a self-described "computer nerd," called ahead and arranged to have three uniformed officers waiting for them at Security. On the way there, McDaniel and Ferguson discussed options, wondering how to detain the woman once they found her. If it were Piper flying under Tina's name, was it a federal crime to fly under an assumed name? If so, maybe they'd have a reason, one that would hold up in court, to take her into custody.

That afternoon in Houston, Tina Rountree called Doug McCann, the man she'd spent the previous night with, and asked if she could come to his house to take a nap. It seemed an odd request, but Doug didn't question it. "Sure," he said. She arrived a short time later and went upstairs, to the third-floor bedroom, and lay down. After Tina left, Doug noticed a message on his cell phone. When he checked it, it was Piper's voice. She sounded upset, and she said she was at Tina's house, locked out, that she'd left her cell phone inside.

"I'm looking for a key," Piper said.

Doug picked up the telephone and called Tina. "Piper's looking for you," he told her.

Meanwhile, that afternoon, Charles Tooke had been at home in his apartment on Lake Conroe, north of Houston, unaware of the drama unfolding in Richmond and Houston involving his friend Piper. At 3:21 that afternoon, his computer popped up an e-mail from Piper's e-mail address: "I have a brief question for you."

Three minutes later, Charles, chuckling, e-mailed back: "I'm a boxer shorts. Not briefs."

He waited, but she didn't reply.

Then, a little more than an hour later, at 4:27 P.M., just as McDaniel and Ferguson were arriving at the airport gate, Tooke received a voice mail from Piper on his cell telephone. "I've got a question for you. Oops, my kids are on the other line," then the phone went dead. Charles tried to call her back. She didn't answer. He left a message, but she didn't respond.

Once inside the airport lobby, McDaniel and Ferguson flashed their badges, hooked up with the uniformed officers assigned to assist them, and headed toward the gate where the Southwest flight in question was scheduled to arrive in minutes. As they rushed to the gate, McDaniel called Stem in Virginia again, looking for more direction. Under Texas law, he'd confirmed, he could stop the woman and finger-print her, to determine who she was. Did they want anything else? Stem called back minutes later. By then he'd talked to Kizer, and they'd agreed that they wanted Tina or Piper traveling as Tina intercepted and detained long enough to confirm her identity.

Once the gates opened, passengers poured out of the

airplane, eighty in all, and McDaniel and the other officers worked the crowd, looking for the faces of the women in the blown-up driver's license photos. They searched, staring into eyes, looking for anyone who avoided their gaze and acted suspicious. As the passengers fanned out into the airport, one officer ran to check the bathrooms, while McDaniel and Ferguson continued the search at the gate. Still, no one looking like Tina or Piper Rountree disembarked. Once the stream of passengers stopped, the two officers entered the airplane, checking the inside, looking for her hiding behind a seat or in the bathroom. It was empty.

They discussed impounding the airplane to give the Houston P.D.'s crime scene experts time to search it for evidence. Perhaps they'd find DNA from a hair left onboard. But with Southwest's policy of no assigned seating, they had no idea where the woman traveling on the Tina Rountree ticket sat. In the end, they decided against it, knowing that it would cost a fortune to process an entire commercial airplane.

Back in the terminal, at the gate, McDaniel cornered an agent who checked the records and discovered that the woman they were after had checked baggage. With that, the two officers ran to the baggage area, only to find the conveyor belt empty and the area deserted.

"That's it," McDaniel said. "We missed her."

"We missed her," McDaniel's boss, Lieutenant Rick Maxey, told Captain Stem when he called that afternoon. The Houston P.D. did, however, have another piece of evidence. Before they left the airport, they'd stayed while the baggage clerk went through the tags. They had the tag reading Tina Rountree, showing that a woman using that identification had checked a bag on the flight. Maxey had the ticket in a plastic evidence bag, so it could be fingerprinted. Meanwhile, Breck McDaniel was on the way to Tina Rountree's house, to see if she could be found there, while Maxey and Ferguson were driving back out to Kingwood to again check Piper Rountree's house.

Grateful for all they'd done, Stem thanked Maxey for the update, but he hung up the telephone disappointed. Now they'd have a difficult time proving who traveled on the ticket. At the very least, someday a defense attorney would have wiggle room to inject reasonable doubt into a trial, by claiming they didn't know who the woman on the airplane had been.

"Heck," Stem said. "Now that's what I call a bad break."

When Stem called Kizer at home to tell him the bad news, the prosecutor paused for just a minute, then said, "I bet she's wearing a disguise."

"Could be," Stem said. "What I do know is we need to get someone to Houston quick to follow this up." When Stem hung up with Kizer, he dialed his boss's phone number, the Henrico County Chief of Police Henry Stanley. "I need your credit card," he told him. "We need an officer on the first flight out to Houston in the morning."

Stanley didn't hesitate, rattling off his credit card number to Stem.

From that point on, the groundwork was laid: Kelley and his partner, Investigator Robin Dorton, a forty-something, soft-spoken officer with a slight paunch and a disarming, Columbo-like approach to questioning suspects, would fly to Houston the next morning to follow leads in Texas, including, they hoped, a sit-down with the prime suspect, Piper Rountree. At the same time in Henrico, Williamson would continue to collect phone and bank records, while Hanna followed any Richmond-area leads.

As for the prosecutors, Kizer had made his intentions clear: If Captain Stem and his crew wanted to charge Piper

Rountree, the Henrico P.D. needed more than conjecture and cell phone records; Kizer wanted nothing less than an eyewitness who could place not just Piper Rountree's cell phone, but the woman herself in the Richmond area at the time of the murder.

In Houston that evening at 6:00 P.M., Breck McDaniel arrived at Tina's house and found Piper Rountree's Jeep Liberty in the driveway. McDaniel called his boss, Lieutenant Maxey. "What does Virginia want me to do?" he asked. Maxey said he'd check.

Fifteen minutes later McDaniel still didn't have an answer from Virginia when Piper walked out the door wearing a tank top and jeans and climbed in her black Jeep, license number X54-JBJ. He followed, as she drove from the house the few blocks to Tina's clinic, then got out of the car. McDaniel parked his unmarked car a few houses away, then walked toward the clinic. He ducked into the adjacent property's backyard, where he had a view of the clinic parking lot. The clinic was closed, the lot empty, except for Piper's Jeep parked next to the trash bin with the car doors open and country music blaring. McDaniel watched but never saw her leave the car or enter the clinic. Inside the clinic, the lights never went on. At seven she pulled out of the driveway, and he ran to his car to follow. He stayed on her tail until she pulled through an amber light and he was cut off by another car. At that point he lost her.

When Breck reached Coby, the Richmond officer told him to let it go for the rest of the night. "Piper is a suspect," he said. "But I'll be there tomorrow, early, and we'll take it from there."

Around the time Piper was in the clinic lot, she returned Loni's phone calls. Loni, by then nearly frantic, would later remember blurting the news out, crying, "Fred's been killed."

Loni waited for a reply, but at first Piper was silent. When she did talk, she said only, "Uh-huh."

"No, you don't understand," Loni said, sobbing. "Fred was shot and killed in the driveway, at the house."

"Uh-huh," Piper said again. "So where are my kids?"

Later, Piper's tone of voice would resonate with Loni. She sounded cold, disinterested, and not surprised.

"So where are my kids? I want my kids," Piper asked again.

"I don't know," Loni said. "The police have them."

After 7:00 P.M., Coby Kelley received a voice mail on his office phone from Piper Rountree: "Hi. I didn't catch your name. I talked with Loni, and I'm looking for my kids. I'm at my sister's." She then repeated her cell phone number ending in 7878, the same one that had been traced to Richmond just that morning.

In Richmond that Saturday evening, Chuck Hanna was making the rounds, trying to find out where Piper Rountree had stayed, assuming she, in fact, was the one who had murdered Fred Jablin. By then he knew that at 4:46 that morning, one hour and forty-five minutes before the murder, someone had checked messages on Piper's cell phone. The call was initiated a short drive from the Jablin house, bouncing off a tower near Cox Road and Broad Street, an area peppered with small hotels. Patiently, Hanna went from one to the next, showing a copy of the photo of Piper in the red dress they'd taken from her son's bedroom, and asking the clerk on duty to check the hotel records from the night before, to determine if a Piper or Tina Rountree had registered.

Late that evening he pulled up into the parking lot of the Homestead Suites, tucked back off the road, hidden behind a shopping center and a line of trees, not visible from the main street. Inside, Hanna went through that same routine.

This time, however, the clerk checked the records and didn't shake her head no.

"A Tina Rountree had a reservation here for last night," the woman said. "But we have no record of her ever checking in."

"You sure she wasn't here?" Hanna asked.

"She never registered," the clerk insisted. "She didn't stay here."

Hanna reluctantly left, moving on to the next hotel.

Back and forth that night, Kelley and Piper traded voice mails. Always, she asked about her children. Finally, at 9:00 P.M., David Ferguson had Piper's house staked out. She wasn't home. That was when Kelley reached Piper on her cell phone.

"This is Piper Rountree," she said calmly.

A friend, Piper said, had told her Fred was dead and the kids were being held somewhere by the police.

"I'm sorry you had to learn on the phone like that . . . your kids are fine," Kelley assured her.

"Well, where are they?"

"That's what I want to talk to you about," Kelley said.

"Loni told me Fred was shot . . . what happened?"

Not wanting to give away too many details, Kelley explained in broad strokes, then asked about their divorce. Fred, Piper said, was physically abusive. "It has not been amicable," she told him. The last time she'd been in Virginia, she told him, was the camping trip. "Can you just tell me where the kids are? . . . I need to come get my kids . . . They're my kids. I have custody of the kids."

"Do you happen to have a copy of that order?" Kelley asked. "I'm just trying to get information on what problems [Fred] might have been having . . . can we talk about that?"

". . . but how that affects the kids, I can't say," Piper answered. No matter what Kelley asked, there was only one thing Piper Rountree wanted to discuss—getting her children.

Then Kelley asked her a question that could help locate her on Friday, if she had been the one who talked with Paxton on Friday afternoon. "I called him yesterday," she confirmed. "I need my kids . . . I'm their mother . . . They're my kids. I have custody."

Now Kelley had her admission on tape, that she'd been the one who'd talked with Paxton. "You understand my situation . . ." Kelley said, indicating that Fred had made accommodations for the children that might prevent them from being immediately turned over to her, including a will that stated he wanted them to be with his brother, Michael.

"I am the mother," Piper said, furious. Then after considering what she'd just heard, she said, "Well, he may have. He's also expressed that my sister, Tina . . ."

"Okay, I was not aware of that," Kelley said. "Is there somewhere where I might be able to speak with her?"

With Kelley's request to talk to Tina, Piper's cell phone went dead. He called back, but Piper's voice mail picked up, and Kelley left a message.

After he hung up, Kelley called the number he had for Tina. She answered, listened to him, then said she was on her way to have a drink with Piper and she'd ask her to call him. Reluctantly, Kelley hung up the telephone. There was nothing more he could do until morning when he landed in Houston. Fifteen hours after he'd arrived on Hearthglow Lane, Coby Kelley was finished for the night.

Much of what had happened that day didn't make sense to Doug McCann. There were Piper's phone calls, and Tina's coming to his house for a nap. He hadn't asked, but he'd wondered how Tina had finished work and driven to Clear Lake to a baby shower as she'd said she was going to, in the rain, and returned so quickly. From four-thirty to six-thirty or so she was asleep in his upstairs bedroom. She didn't seem upset, just tired. He'd expected that he, Tina, Piper, and Jerry would be going out to dinner. After all, that was what Piper had mentioned in the morning, but through the evening, his phone didn't ring. No plans were made. By eight-thirty he was getting hungry. He called Tina, but she sounded busy and said she'd have to call him back. When she finally called, she said, "Something serious has come up. I can't talk about it now."

In Baton Rouge, just after eleven, Jerry Walters's phone rang. He'd had a rough couple of days. Bertha, his beloved bloodhound, had surgery on Friday and he'd visited her at the vet's office on Saturday. The dog wasn't doing well. When he answered the telephone, it was Piper, and she said, "Fred is dead."

Fred, she said, had been murdered in the driveway.

"Can you come to Houston to be with me?" she asked.

Jerry felt unsure. "Let me see what I can do," he said.

It was after midnight when the telephone rang at Doug McCann's town house. She'd sounded upset earlier, but now Tina was upbeat. "Can I ask a favor?" she asked.

"Sure," he said.

"Is it okay if my sister comes to your place with me tonight?"

When he agreed, saying Piper could sleep on the downstairs couch, Tina said they'd be there soon. Still, it was nearly 2:00 A.M. when Tina and Piper arrived. Doug heard them come in, and then Tina walked up the stairs to the bedroom. "I'll tell you about it tomorrow," she whispered as she climbed into bed. "I don't want to talk about it tonight."

Doug and Tina lay there briefly, in silence, before Piper walked into the room and lay on top of the blanket and between them at the foot of the bed.

"Piper, how are you?" Doug asked.

Piper mumbled something Doug didn't catch.

Just then, Tina sat up and reached down to tenderly hug her. "I just love my little sister," she said.

● ●

11

Captain Stem often marveled at the advances science had brought to modern criminal investigations, from DNA to microscopic fiber analysis. It was such technology that yielded an important break on day one of the Jablin murder investigation: the trail Piper Rountree's cell phone left as it traveled through Richmond. That knowledge had led the police to the specific Southwest Airlines flights a woman using the name Tina Rountree had taken from Norfolk to Baltimore then Houston. Then police suffered their first setback, when the Houston P.D. was unable to intercept the woman at the airport.

On day two of the investigation, while they'd continue to use everything modern technology had to offer, Stem knew it was time to rely more heavily on a traditional police tool—old-fashioned gumshoeing, hitting the pavement to track down witnesses and collect evidence. In Richmond, he assigned Investigator Chuck Hanna to follow leads, while Coby Kelley and his partner, Robin Dorton, were on an airplane headed to Houston to pick up Piper Rountree's trail.

It was that need to gumshoe the case that brought Hanna to his desk in the low-slung, brown brick building that comprised Henrico P.D. headquarters at nine that Sunday morning, Halloween day. Once there, he made copies of the photo of Piper Rountree confiscated from Paxton's bedroom, the one of Piper in a red dress. He then got in his unmarked car and returned to the area around Cox Road and Broad Street, where Sprint records showed Piper's cell phone had been used two hours before the murder. His assignment: to find what Wade Kizer said was essential to the case—a witness who could place not only Rountree's telephone but the woman herself in Richmond on the day of the murder.

Yet two hours later, after questioning employees in most of the restaurants in the area around the Cox Road tower, showing them Rountree's photo, Hanna had come up empty.

"No, haven't seen her," he heard over and over again.

That left the hotels.

As his first piece of business, Hanna returned to the Homestead Suites, where he'd learned the previous evening that Tina Rountree had an unused reservation. Again he talked to the desk clerk and all the personnel on duty, showing them Rountree's photo. Again no one remembered the woman in the red dress. Disappointed, Hanna asked the desk clerk to double-check the records by printing a complete copy of the hotel register for the nights of October 28 and 29, the two nights the Southwest Airline tickets suggested Rountree was in Richmond. To Hanna's disappointment, he still didn't find the name Rountree on the list.

With that, he left, vowing to keep trying until he'd inquired at every hotel in that part of the city.

Meanwhile, in Houston, Charles Tooke had been trying to reach Piper since Thursday. He needed paperwork from her, something for work. He'd called repeatedly, but she hadn't answered her cell phone all day Thursday. On Friday, Charles finally reached Mac, who told him that Piper had said she'd be in Fort Worth on Thursday, Friday, and Saturday, attending a law conference. So at 9:39 Sunday morning, Charles tried again, dialing Piper's cell phone, the 7878 number he'd always used to reach her in the past. This time, Piper answered.

"Fred is dead," she blurted out.

The conversation was short that morning. Piper, who sounded sad and tired, never explained to Charles how Fred died, just that he'd been "killed." With no reason to assume otherwise, Charles pictured a car accident, not a murder. Concerned about a woman he'd grown to view as not only a coworker but a friend, he offered to help if he could, but Piper said there was nothing he could do. After they hung up, he dialed Tina's phone number, wanting to make sure she knew.

"Yes, I know," Tina said, sounding not the least upset.

"How did it happen?" he asked.

"I don't know," she said. "But Richmond is a very violent town."

That same Sunday morning, Breck McDaniel and his partner, David Ferguson, were busy as well. At Hobby Airport, Ferguson waited for Coby Kelley and Robin Dorton's plane to arrive. After Ferguson picked them up, he was to drive them directly to rendezvous with McDaniel, who had Piper's Kingwood house staked out. Just after the flight landed, McDaniel called Ferguson to tell him that Piper Rountree had just driven into her garage in her black Jeep Liberty, license X54-JAJ. She'd closed the garage door and presumably gone inside.

"What does Kelley want me to do?" McDaniel asked.

"Hang tight until we get there," Ferguson answered, passing on a message from Kelley. "We're on our way."

Half an hour later a Houston officer McDaniel had made arrangements to have on-site as backup relieved him of his position watching the house, and McDaniel drove to the end of the block, where he met with Kelley, Dorton, and Ferguson.

"Time to knock on her door," Kelley said with a smile.

This time when Ferguson and McDaniel drove up in their

unmarked cars, they didn't park down the street but pulled up directly in front of Piper's house. All four officers walked to the front door, and Kelly pounded solidly. No answer. He rang the bell. Again no answer.

Making sure she hadn't left, Ferguson walked around to the garage and peeked in the window. As he expected, Piper's SUV was inside. He returned to the front door and the officers rang the bell and knocked again. Still no answer. They knocked harder, but without a response. Pulling out his cell, Coby dialed Piper's house phone. Within seconds they heard the phone ringing inside the house, but no one picked up. When the voice mail came on, Coby left a message: "Ms. Rountree, this is Investigator Coby Kelley from Henrico County. I'm outside your door with other officers and we'd like to talk with you. Please come to the door." He hung up and they waited, but the door never opened.

Through the windows, all four officers saw what appeared to be an orderly home. More than one noted a computer tower in the living room underneath a desk, clearly visible from the front door. Breck made a note in his pad: "Desktop computer, living room." That was something they'd list when they sat down to write up a request for a search warrant.

Even after Breck McDaniel walked around the house and banged on a small bathroom window in the back, no one came to the door to let them in.

"What do you want to do?" Breck asked Kelley. "Should we try to get a search warrant now?"

On his cell phone, Coby called Sergeant Russell in Richmond. When he hung up, Kelley said, "They don't want to do that yet. Let's leave Ferguson here to watch for her. The rest of us can try to hook up with Tina. I'd like to talk to her before the women are able to compare notes on what we're asking."

With that, Dorton, Kelley, and McDaniel left to drive back

into Houston, hoping to track down Tina Rountree, while Ferguson settled in, watching Piper's house. After the others had left, Ferguson took out his notepad and wrote down what they'd done so far. Such notes could become important evidence to put before a jury, documenting a suspect's odd behavior—such as refusing to answer a door when police were trying to talk with them.

It was then that Ferguson saw the black Jeep Liberty suddenly back out of the garage and down the driveway. Inside the SUV he saw a woman resembling Piper Rountree, who hit the gas peddle, the Jeep skidding off down the street.

Quickly, Ferguson started his car and pulled out, falling in behind her.

"She's on the move," Ferguson said to Breck on the telephone.

"We're on our way," Breck answered.

Taking the next exit off the freeway, McDaniel swung the car into a quick U-turn and headed back toward Kingwood.

Minutes later Ferguson called his partner again, giving them his location: "Looks like she's pulling into a strip center parking lot."

"Tell her we're here from Virginia and we want to talk with her," Kelley told Ferguson. "Try to keep her talking until we get there."

In the parking lot, the Jeep swung into a space in front of a PetSmart store. Hurriedly, Ferguson pulled in beside her, parked and jumped out. "Ms. Rountree," he called as Piper, dressed in cutoffs and a T-shirt, bustled toward the pet store. "I'm Sergeant David Ferguson, and I need to talk with you," he said, flipping open his badge for her to see. As instructed, he then launched into the explanation: Kelley and Dorton were in Houston from Henrico and needed to meet with her, and would she mind just waiting a few minutes while they drove to the parking lot. Or could they make other arrangements, something a bit more comfortable?

"They need to work out getting your kids to you," Ferguson said, dangling the bait. "Can we go back and talk at your house?"

Given the option, Ferguson wanted Piper in her house, so he and the other officers had the opportunity to convince her to sign a consent-to-search form. That would make things easier, since they'd no longer need to ask a judge to sign a search warrant.

Yet, it appeared Piper had other things on her mind.

"No, if you want to talk, we can talk here. But first I need to run an errand," she said. "I have to get crickets to feed my frogs."

Amazed that she didn't seem the least bit worried about her children or concerned about what he had to tell her, just a day after her ex-husband's murder, Ferguson pressed, "So, after you buy the crickets, you'll have time to talk to us? Maybe we can head back to your house there, talk where we can have a little privacy?"

"I'll be right back," Piper said, leaving him standing in the parking lot as she walked into the pet store.

After Piper disappeared inside, Ferguson called McDaniel to fill him in. The elation of finally cornering her, however, evaporated just minutes later when she emerged from the store and hurried past him, back toward her SUV, not even looking at him.

"The Virginia officers are on their way, and I can have them meet us at your house to talk," Ferguson said again, smiling.

"No," Piper said. "The Virginia police haven't even contacted me. If they want to talk, tell them to call me."

Ferguson knew that wasn't true but didn't argue the point. "Ms. Rountree, let's talk about this," he said, but Piper swung open the door to her Jeep. She slammed the door, quickly turned on the engine, threw it into reverse and backed up, then revved forward out of the parking lot. By then Ferguson had jumped in his own car, and he followed her into traffic.

"She's on the move," he told the others via phone. "Looks like she's headed into Houston."

McDaniel, who'd driven halfway back to Kingwood, pulled off of Interstate 59 again and made another U-turn. This time he sat and waited. When Ferguson called in a position two exits north of where he'd parked, McDaniel, with Kelley and Dorton in the car, pulled back onto the interstate and headed slowly south, toward downtown. When McDaniel saw Piper's Jeep followed by his partner's unmarked car, he pulled alongside them. Ferguson opened up enough room to allow McDaniel to cut in front of him, and they formed a three-car caravan, with Ferguson's as the third car. McDaniel knew that with all the jockeying for position they'd been too obvious for her to ignore, that Piper must have known she was being followed by two police cars. But she didn't pull over. Instead, she drove through Houston, past the city's forest of mirrored skyscrapers, their top floors obscured by low clouds, and kept going. At first it appeared she was heading to Tina's house or the clinic, but instead she pulled up and parked in front of a red brick duplex. Out front was a sign that read: ATTORNEYS, MARTIN MCVEY. Still visible was the name Piper Rountree, under a coat of silver paint McVey had applied after he'd ordered her to move out.

As Piper stepped from her Jeep, Kelley was only a stride behind.

"Ms. Rountree, I'm Investigator Coby Kelley from Henrico," he called out to her as he followed her toward McVey's front porch. "I talked with you last night, and I'd like to talk to you again. We have a few questions, and we need to discuss the situation with your children."

Piper turned, looked at him and said, "Come in."

Inside McVey's homey first-floor office, Kelley entered a blue-walled reception area with book-lined shelves and burgundy leather furniture. Off to the right, in a side room, Kelley saw a blond, attractive woman, Tina Rountree, sitting across the desk from a large man with white hair and a beard, Marty McVey.

When she noticed Coby, Tina smiled and asked, "Who's that handsome devil coming in the door?"

Smiling, Kelley introduced himself, but Tina's demeanor immediately flipped a full 180 degrees. Suddenly crying, she ran from the room, toward the back of the house. Not forgetting who he'd followed in the door, Kelley turned his attention on Piper.

At first they talked while Piper paced about the room, at times sobbing. "I know you're really concerned about your kids. I want you to know that they're okay," Kelley assured her. "They're fine. They're with friends."

Before he'd left Virginia, Kelley had called Officer Boyd, the neighbor who'd taken the Jablin children in. Kelley had explained that Piper was "all about her children," that she kept asking where they were and wanting to see them, to claim them now that their father was dead. When he got to Houston, Kelley wanted to be able to put Piper on the telephone with her children, to assure her that they were safe and get rid of that objection. Boyd had agreed to put the call through. Now, Kelley launched his plan into action.

"I know you want to talk to your kids, so let's get them on the telephone right now," he said to Piper, dialing the Boyds' house. The timing, however, didn't work. Michael and Elizabeth Jablin had arrived, and she had taken all three children to a neighborhood Halloween barbecue and hay ride at the cul de sac down the street from their home, still ringed in crime scene tape. The children had planned to go with Fred that day, and asked to see their friends. Later, Mel would remember her daughter Chelsea and Jocelyn sitting and painting the other children's faces, Jocelyn looking dazed.

Disappointed, Kelley hung up. They had to wait, he explained to Piper, until the children called back.

"While we're waiting, do you have any questions about what happened to your ex-husband?" Kelley asked.

Piper replied, "No, Loni explained everything to me."

"Where were you this weekend?" Kelley asked, getting right to the point.

Piper turned her eyes from him and looked visibly uncomfortable. He'd later describe it as "squirming."

"Well, ummm . . ." she said.

As Kelley started to press her, the phone rang. It was Harry Boyd with the Jablin children, who'd been retrieved from the Halloween party, ready to talk to their mother. Kelley handed Piper the telephone.

For the next thirty minutes Piper talked to her children. While she talked, McVey introduced himself, telling the four officers that he was a former Harris County prosecutor and a criminal and civil lawyer. Kelley listened, but kept one ear on Piper's conversation, hearing her counsel the children, as if they were her clients: "You don't have to talk to them. You have the right to have a lawyer."

At one point Piper was heard telling the children to "hang tight until you see a signed custody order," then, "Paxton, your dad is dead. I don't want you to be next . . . Paxton, I'm afraid for you guys . . . Your uncle Mike will inherit everything when you die. Paxton, there's probably millions. If I were you, I'd ask for police protection . . . I'd prefer you guys be with Loni."

To Callie, Piper said, "I'm trying to figure out why you can't come down here."

A little while later Kelley heard her advising one of the children to be suspicious of Michael Jablin. "Your uncle is the only one who would profit from this," she said.

Kelley listened, making a mental note of what was being

said, the entire time carrying on a friendly conversation with McVey and the other officers. After she finished, Piper handed Coby his cell phone, and the officers guided her over to the reception area. Coby sat on the couch, next to Piper, who looked small and uncomfortable in the high-backed, tufted leather chair. McVey went back to his desk in the next room, sitting a dozen steps away, listening. He'd say later that he felt as if he were being drawn into a cyclone.

"Listen, I overheard your conversation. You seem to be indicating that Michael Jablin or the University of Richmond may have a motive," Kelley said to Piper.

"Yes," she answered, wringing her hands and twisting a white tissue until it resembled a thin white cord.

Just then Breck McDaniel surreptitiously turned on the small digital recorder he'd hidden in his shirt pocket.

"Are you saying that you think Michael or the university is involved?" Kelley asked.

"I don't know," Piper said, her voice breaking into violent sobs. "First of all, please don't let them go with . . ."

"Mike?"

"Yes," she said, crying even harder.

From his office, McVey interjected. "She's going to tell you some things that were going on with her ex-husband's life and stuff like that. She's afraid that you folks are going to turn around and tell the kids."

"Okay," Kelley replied, calmly. ". . . We are not about trying to taint anyone's view of their father . . . my interest is in pursuing this criminal case . . . in making sure the kids are safe and in a good environment. I don't know her, and I certainly didn't know her ex-husband. If we're about to place [the children] in harm's way, we certainly need to know that."

"The only people who have anything to gain monetarily by Fred's death, as far as I know, is Fred's brother and his wife," Piper said. The brothers, she claimed, were far from close, rarely seeing each other or talking. "Michael has access. He's the beneficiary, I'm sure, on Fred's estate."

As they talked, Piper threw out accusations, for which she offered no evidence. Perhaps, she suggested, her former brother-in-law had murdered Fred to get control of the estate and Fred's life insurance money. While under Fred's will the funds were to be kept in trust for the children, she claimed, "Michael would have control of their monies."

Piper drifted off, her mind carrying her into tangents, complaining about Michael as if he were Fred, charging that he'd kept the children from her, that he'd been the one who refused to give the children money to visit her in Houston. "He kept us apart," she sobbed.

She then turned her attention to the University of Richmond. Inflating UR's more than $1 billion endowment to $3 billion, she claimed that Fred "had control of it all." Not only did he have control of all the university's funds, but, according to Piper, Fred Jablin, a professor who didn't even sit on the school's board, also controlled all the people who worked at UR. As Piper described him, Fred was a powerful and secretive man, one who had a private room in the garage, which he kept locked behind a steel door.

"I didn't understand what was going on in the marriage," she said methodically, ripping the white cord she'd formed into small shards of tissue. "And, my sister deals with women's issues . . ."

"Your sister, Tina?"

"I know, she's a flake," Piper said, her voice suddenly controlled and devoid of sadness, instead ironic with a slight laugh. "But . . . Fred for years grew marijuana and sold it. Used it all the time . . . He took trips. Delivered things."

For twenty minutes Piper threw out accusations, implying Fred wasn't just a professor but a big-time drug dealer, and that any one of the people in his life but she could be responsible for his murder. Fred, against her wishes, she said, had kept a gun in the house, on a shelf. Perhaps, she insinuated, that could have been the murder weapon.

Attentive, Kelley was supportive, pushing gently, trying to build a rapport with Piper, and hoping she'd open up. He listened and took notes. What he needed from her, she wasn't offering, the answer to a simple question: Where had she been the previous morning, at the moment Fred Jablin was murdered?

That afternoon with Kelley and the other officers, Piper again claimed Fred had abused her. When Kelley described the divorce as acrimonious, Piper blamed it on the courts, saying Fred had parlayed his influence and claiming that he and his attorney had held meetings with the judge when she and her attorney weren't present, never explaining that she'd been notified of the hearings and simply hadn't shown up.

Although just weeks earlier Fred had let the children go camping with her, Piper then said, "The kids were afraid they'd never see me again, and I was afraid they'd never see me again."

Kelley listened, nodded, sympathized, but eventually he worked the conversation around to the question at hand: "Did you go to Virginia at all this weekend?"

"No," Piper answered.

The entire time she talked, Breck McDaniel watched Piper Rountree. After years of interviewing suspects, he considered himself something of a human lie detector, and he had no doubt that this woman was lying. She was evasive, slow to answer, and didn't appear at all as concerned with her children as she claimed to be.

"Can you just let us know where you were and where somebody might have seen you, so we can knock that out and kind of eliminate you as a suspect," Kelley asked.

"Uhhh . . . right here," she answered.

"In this room, in this house?"

"No . . . in Houston and Galveston."

"Were you working?"

"I work all over."

As Kelley pressed, Piper hemmed and hawed, not giving any clear answers. She thought she'd been with Mac, training him. "Yeah, that was it," she said.

"Where were you when you talked to Paxton?" Kelley asked.

"I was in the car driving back from Galveston," she said, calmer.

That was an important admission, since that call had already been traced to a Richmond cell phone tower. When Kelley asked about her cell phone, trying to get her to recite the number the call had been made on to confirm she'd been the one on the 7878 number the day before the murder, however, Piper said she couldn't remember her cell phone number.

"I'm trying to help you reflect back," Kelley said calmly, but with determination. In order to eliminate her as a suspect, he had to be able to verify where she was at the time of the murder, he explained again. "We just want to establish where you were Friday through Saturday, because this happened early Saturday morning." Kelley asked again where she was on those two very important days. Again Piper didn't answer. He persisted.

Finally, she implied she'd been with a married man: "Let me tell you this: I don't want to tell you about the person because it involved somebody else. I mean, as far as relationships, it's nothing to do with this."

"You understand this is a homicide investigation?" Kelley said, his voice becoming louder and firmer. "If ever there was a time to come to Jesus—"

"I'm at peace with Jesus, okay?" Piper bristled, her voice strained and high-pitched.

"I don't care if you were buying cocaine from a

minister . . . you know what I'm saying," Kelley said, even more insistent.

Moments passed.

"I was at Tina's, but she wasn't there," Piper offered, suddenly calmer. She continued, claiming she'd seen Tina sometime that Friday afternoon, as her sister went back and forth between the house and her office.

"Obviously, we'd rather have a priest you were with, she's your sister, but okay," Kelley said. "And what about Friday night?"

"Uh, hmmm," she agreed, indicating she'd been at Tina's that night as well. But then Piper backtracked. At first she said Tina hadn't spent the night at the house, then that perhaps her sister had been there.

"It's a big house," Piper said. "I don't know when people come and go."

Kelly reiterated what was forming as Piper's alibi: that she'd been at Tina's house Friday afternoon and night. Would her sister confirm that?

Again Piper balked.

"Can we just leave it at that? My concern," Piper said, measured, angry, and biting off her words. "I have my children I'm concerned about. It may sound cold, but something has happened, and now you're ready to turn the kids over to him . . ."

With that, she flipped full circle, back to Michael Jablin, who she described as "not a trustworthy person," and certainly no one who should have custody of the children. "I had no way of knowing that Fred had signed his life over to him."

"You say you were in Texas, you weren't at all in Virginia," Kelley said, bringing her back to the most important question, her alibi. "Tell me where you were, when you were with Tina, and I'll try to put a hold on what's going on in Virginia."

McDaniel piped in, "Somebody can verify your whereabouts?"

"Yeah, that's what I'm looking for," she said.

At that point, from his office, McVey rejoined the conversation, urging Piper to give the officers what they needed, to warn the man she'd been with before the police banged on his door.

"We're not going to put his business out on the street," Ferguson tried to reassure her.

With that, McVey, who had to pick up his teenage son, Alex, suggested they call the meeting to an end and meet again the following morning, after she'd had time to "tell the man what was going down."

Piper agreed.

Coby walked off to use his cell phone, but Piper kept talking, as the other three officers attempted to get her to verify key points in what she'd already told them.

"You talked to Paxton on your way back [from Galveston]?" Ferguson queried.

"I talk to Paxton all the time," Piper answered, sounding irritated. Again, when Ferguson asked, she said she couldn't remember her cell phone number and didn't know where her cell phone was over the time of the murder, at Tina's house or her house.

"These investigators need to know your whereabouts for Friday and Saturday," Ferguson said, gruffer and more demanding than Kelley had been. "This is for your benefit."

"I understand that . . ."

"You've got to talk to this guy and that's it," McVey urged her, sounding as if he were talking to a little girl about the need to ingest a bitter medicine in order to improve her health. "You're going to have to give up his name. He's going to call them and do it the easy way, or it's the hard way . . ."

"Somebody who lives here in Texas?" Ferguson asked.

"Well, there are two people," Piper said.

By the time Kelley returned to the conversation, Piper had agreed she'd call the man who she said would be her alibi, and then meet with them that very night to give them the information that could answer all their questions—the details on where she was and who she was with at the time of the murder.

Then Kelley, still trying to pin her down, asked how to call her, repeating her cell phone number, the one used in Virginia that ended in 7878. Piper stammered, saying that was one of her phones but she had others.

Again, as they got ready to part, Kelley dangled the carrot that he knew meant the most to Piper Rountree—her children. "I was just talking to the commonwealth attorney," he said. "He's trying to work something out with CPS."

Later, outside, as the traffic streamed by, Piper confirmed something Kelley had wondered about: that she knew her ex-husband was dating.

"Have you talked to Fred's girlfriend?" Piper asked. In Piper's version, the woman Fred had just met via the Internet that spring had been dating him for nearly five years, two years before the divorce. Fred, she claimed to the officers, had had many extramarital affairs. Portraying herself as a dedicated mother, she said she'd been busy caring for their children and hadn't been aware of them at the time.

Sounding sympathetic, Kelley said many women "bury their heads in the sand."

When they parted, Piper agreed to call Kelley at six that evening, to talk about what he had been able to arrange in Richmond to keep the children from being turned over to Michael Jablin. At that time, she said, she would give him the name of the unidentified married man who could confirm she was in Houston during the murder.

"Like I said, you've got my phone number," Kelley said, drawing the interview to a close. "All right. Thanks."

Finally, ninety minutes after it began, the interview ended, and Piper walked back toward McVey's office, while Kelley and the others drove the short distance to Tina's house. During the heat of the discussion with Piper, Kelley had asked McVey where Tina was. The officers had not seen her leave his offices, but McVey maintained that she had, that she'd walked through the waiting room and out the door while they were all talking.

When Kelley knocked on Tina's door, Mac, home from the Hill Country and a weekend with his parents, answered. He'd learned about Fred's death from Tina early that morning, before he drove in to report to the radio station where he was working. After they showed their badges and introduced themselves, the officers asked to talk to Tina.

"She's not here," he said. "She's down the street, but I'll call her."

Mac picked up the phone, dialed a number and talked to someone, then said, "I'll go pick her up. You're welcome to stay. I'll be right back."

With that, he left. Later, Mac would tell Kelley that where he went that afternoon was right down the block, to McVey's office, where he found Tina and Piper talking. Minutes later Mac returned with Tina, who was immediately on the offensive.

"Why aren't the children here?" she demanded. "They should be with their mother."

"We need to talk to you about your sister," Kelley said. "Did you see her this past Friday or Saturday?"

At first Tina was evasive, continually asking about the children, pushing for Kelley to say he would bring them to Piper or make arrangements for her to pick them up. But Tina wouldn't commit herself as to when she saw Piper over that weekend. Finally, she said, "Yes, I know where my sister was this weekend. You bring the kids here and I'll tell you."

Back and forth they went, Kelley, Dorton, Ferguson, and

McDaniel, all pushing Tina to tell them where Piper had been that weekend, but she wouldn't comply. "You leave and come back with my sister's children," Tina ordered. "Then we'll talk."

She complained about the Virginia court system, saying she didn't like the way the divorce case had been handled and that she didn't trust anyone in Virginia in power. When Ferguson told her he was from Houston and had nothing to do with the Virginia police, that he'd be happy to take her statement, Tina didn't mince words.

"I'm not answering your questions, either," she said.

"Where were you on Friday night, and did you see Piper?" Kelley asked again. "That's all we're asking."

"I'm *not* answering your questions," she said again.

Finally, as Tina became increasingly hostile, she ordered them to leave her house. At that point Kelley decided they might just as well go without a fight. It was obvious that Tina Rountree wasn't going to make any kind of a statement. They left, walking outside toward their cars. Yet, Tina didn't let them leave in silence. Instead, she followed them out the door and onto the street, shouting, "Bring my nieces and my nephew here. They need their mother."

Kelley thought he could still hear her hollering at them as he drove away.

About one that afternoon, in Virginia, Steve Byrum's phone rang, shortly before the meeting in McVey's office. It was Piper, and she left a message: "I need your help. Call me." When he got the message later that afternoon, Byrum called Hanna.

"Would you be willing to record a phone call with Piper?" Hanna asked.

"I'd rather not," he said. "I don't want to end up like Amber Frey in that Scott Peterson case, pulled in on something I had nothing to do with." After they talked awhile, Byrum relented and agreed. Hanna said he'd make the arrangements to make the call later in the week.

Finally, at four that Sunday, after a full day of circulating from hotel to hotel without finding anyone to identify Piper Rountree, a disappointed Hanna left work and headed to his ex-wife's house to pick up his young daughter. It was Halloween, and they had plans. The youngster had a Barbie princess costume to wear, and the big, muscular officer had every intention of taking his little girl trick or treating. Yet, he couldn't quite put the Jablin case behind him. Throughout the evening, he thought about Fred Jablin, about how he must have been looking forward to Halloween with his three children, putting on costumes, walking through the neighborhood, laughing with friends. Now that would never happen again.

At 3:55 that afternoon in Houston, shortly after Piper finished her first interview with the investigators, the phone rang at Charles Tooke's house. Piper told him she'd been interviewed by the police and that she thought it had gone well. "Can you come over tonight?" she asked.

"Sure," Charles said.

Three minutes later she called a second time.

"You know that piggyback cell phone you have?" she asked.

"Sure," he said.

"Can I use it? Would you bring it tonight?"

"Of course," he said.

The officers' plans to meet with Piper that night evaporated with a 4:27 voice mail from her on Coby Kelley's office phone. When he called her back, she cried, saying she felt caught "between a rock and a hard place."

"Do you feel like you're being accused of something?" Kelley asked.

"I don't know," Piper said.

". . . Can you tell me what makes you uncomfortable so I can work on it, ma'am?"

"I'm just fully exhausted . . . I need to find out, to figure out how to get my kids, custody of the kids in Virginia."

In the end they agreed to meet at McVey's office again the following morning, at nine.

In Richmond that evening, outside the Jepson School of Leadership, in a grassy area called "the quad," a group of thirty or so students and faculty gathered to hold a candlelight vigil for Fred Jablin. After sunset, they stood together, in the cool evening, and recounted stories of their dead professor and friend. One student, a junior, had been so upset when she heard the news that a friend drove her around the campus while she cried, nearly hysterical. That night, she recounted her memories of Fred Jablin, how she'd worked with him on her project just that week. She'd wanted to do a paper on leadership and surfing, and rather than trying to dissuade her, Fred had been intrigued by the concept, grinning ear-to-ear, as he often did when excited about an idea. "This reminds me of my kids," he'd said. "They like to talk about surfing."

The newspaper and television reporters were kept back from the ceremony, as the university's spokesperson, Brian Eckert, stood with the others, holding a shimmering candle and thinking about how it was such an odd turn of events for the university. It seemed strange that something as violent as murder had touched such a civilized place.

How could it happen to a member of our faculty? he wondered.

At a meeting held earlier in the day, Eckert and others had told the students all they knew: just that Professor Jablin had been murdered and no one had yet been arrested. Many of the students cried.

That Sunday, word of Fred's murder traveled quickly through academic circles. At the Houston airport, preparing to board a flight back to Austin after delivering a lecture, John Daly received a call from a colleague to tell him of Fred Jablin's death. The first thought that came to him was: God knows, it couldn't be Piper. Even she couldn't do that to her own kids.

Yet, in Richmond few thought of any other alternative. "We all knew somehow Piper had to be involved," says Professor Ciulla.

At the UR ceremony, as the candles flickered, Professor Elizabeth Fairer and one of Fred's Jewish students recited the Kaddish, the Jewish prayer of mourning.

"He who makes peace in His heights, may He make peace upon us and upon all Israel," they read.

"Amen," the others responded.

"How do I prove where I was without telling them where I really was?" Piper asked Charles after he arrived at her house in Kingwood that evening, just before 6:00 P.M. While the candlelight vigil went on in Virginia, Charles sat in Piper's living room on the beige leather couches he'd helped her buy, discussing how she could establish where she was at the time someone murdered her ex-husband.

As Piper told it, she'd been mistaken about the weekend of the conference and had never gone to Fort Worth. Instead, she was at Tina's, waiting to be picked up to go to a Halloween party with her sister, Jean. However, Piper said, Jean had never shown up. Instead, Jean went to a different party. About 7:00 P.M., Piper contended, she'd given up on Jean and walked down the street from Tina's to a bar called Under the Volcano. She met a married man there, talked to him, and he walked her home at about nine-thirty.

"I think his name was Steve," she said.

As an ex-spouse, of course, she told Charles it was only

natural that she was considered a prime suspect in Fred's murder. While she had nothing to worry about, since she wasn't in Virginia, she didn't want to tell the police about hooking up with the man called "Steve" at the "Volcano."

As Piper talked, she contended that the Virginia court system was against her and couldn't be trusted. She'd been there and done that, and it hadn't ended well. She'd lost her kids, for nothing really, no real reason. If the judge determining custody heard she'd been in a bar, that she'd let a man walk her home, she might never get the children.

She asked over and over again: "How does a person prove where they were on a particular date?"

At times, as they talked, Piper nervously paced the room.

Believing in her innocence and eager to help, Charles had ideas. He knew that Piper rarely cooked. Did she buy wine or carryout? If so, perhaps someone at the store or restaurant could identify her.

"No," Piper said. She hadn't gone out, just back and forth to Tina's house.

"What happened?" Charles asked Piper. "How did your ex die?"

"They said the killer used a gun and some other weapon," she told him. Later she mentioned that police said Fred had been murdered in the driveway. When Charles asked if the children had heard or seen anything, she said they didn't. "They were asleep in the house. It's a big house."

Looking back, Charles would say that Piper didn't seem anxious about how the children were taking their father's murder. In fact, she contended that when she'd talked with them that afternoon, their main concern was whether they'd be going out for Halloween.

At that moment, Piper's primary worry was how to prove she wasn't anywhere near Richmond, Virginia, on either Friday night or Saturday morning. As Piper explained it, she was worried because years earlier, as a prosecutor, she'd seen police mischaracterize statements and shape testimony to frame innocent people. "It was tough to get the convictions overturned, even when there was evidence," she said.

As they sipped wine, they talked.

Charles thought through the situation, advising her that to get custody she was going to have to go to Virginia, hire a lawyer, and fight for her rights.

"I'm not going to Virginia. I'm terrified of the courts there," she said. "I'm terrified of Virginia law. I've been beat up by it before. They'll say I'm an unfit mother." Raising her glass, she added, "Just for having a glass of wine."

As Piper talked about the children, Charles kept reminding her of the ongoing murder investigation. As he saw it, she needed to pay attention to that first, to clear her path to gain custody. Once she was cleared, there'd be time to pursue getting the children.

At one point they discussed the e-mail he'd received from her on Saturday afternoon. That could be used to help prove she was home at her computer when it was sent. Appearing troubled, she mentioned, "There are things in my computer I wouldn't want the police to see."

Charles assumed her worries related to her use of Internet dating services, like the time she used Tina's Match.com account and met Dean Lowry. Despite her concerns, Charles advised her not to touch or change anything on her computer. "Even if you erase it six times, the professionals can reconstruct it," he said. "Anything you change can reflect badly."

When she kept circling back to custody, Charles grew impatient. He was the son of an attorney, and he understood what was at stake, even if she didn't. "Even if you were in an alley with a bunch of men paying them twenty dollars a throw for sex, you tell the police where you were and who you were with," he advised. "You get this murder investigation out of the way, and then worry about custody."

If she couldn't prove where she was, Charles advised, perhaps the forensics on the scene could clear her. "Maybe the ballistics on the gun will match it to a burglary in New Jersey or something," he said.

"How do you match a bullet to a gun?" Piper asked.

She'd told Charles about her time as a prosecutor, and he was surprised at the question, but wrote it off as her being upset at the murder. "Well, you dig the bullet out of the body," he said, but then he had second thoughts, considering that it sounded too cold. "Oh, I'm sorry. He's your former husband."

"No, it's okay," she said, not looking the least bit offended.

Always, Charles turned the conversation back to how Piper could prove to police where she'd been that weekend, and that she wasn't in Virginia murdering her ex-husband. Charles asked if she had any receipts, if she'd bought anything.

"No," she said. "Nothing."

"Did you charge anything on your credit card?" he asked.

"No, my credit card was stolen and my debit card was declined when I tried to use it," she said, adding that she'd gone to the bank to withdraw money to pay cash that weekend.

And then Charles asked Piper if she'd made any cell phone calls over those two days.

"If you used your cell phone, they can trace where you were at the time, based on what tower the signal bounced off of," he said. "As you talk, they can actually follow the cell phone, as it bounces from tower to tower."

Looking back, he'd recall that Piper didn't look particularly worried at that bit of information, but she did say, "Oh."

Moments later she said, "My cell phone was lost. I don't have it."

Charles thought for a minute and then said, "You better hope that your phone and credit card weren't used in Virginia."

"Yeah," Piper answered, laughing slightly.

At nine-thirty that night Charles hugged Piper goodbye. Then he took off the St. Christopher medal he wore on a chain and hung it around her neck.

"You're going to need this," he said.

Fred Jablin came alive in the classroom. He loved nothing more than connecting with a bright student.

Photo courtesy of the University of Richmond, Jepson School of Leadership Studies

Physically, Fred was a slight man with an unassuming manner. Few would have guessed he ranked among the top experts in his field in the world.

Photo courtesy of the University of Richmond, Jepson School of Leadership Studies

Halloween had always been Fred's and Piper's favorite holiday, here at a party at Fred's Austin house.

Photo courtesy of Leo and Linda Kuentz

At first, Piper and Fred seemed happy, content to be together. Here watching football in Austin.

Photo courtesy of Leo and Linda Kuentz

To her therapist, Piper *(far left)* described her older sister Tina *(left)* as her surrogate mother. Everyone who knew them understood how close the sisters were.

Piper, here with Paxton, considered herself the perfect mother. She was beautiful, artistic, and brilliant.

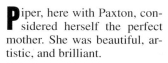

There was little Piper appeared to enjoy more than time with her children, here with Paxton and Callie on the beach.

Photo courtesy of Loni Gosnell

From the beginning, Piper, pictured here at a neighborhood party, stood out from the other women, those she labeled "Stepford wives."

Loni Gosnell was Piper's best friend. "She taught me how to enjoy life," she says. "How to drink my wine out of my best crystal, and not save it for a day that never happened."

Kathryn Casey

From the outside, all appeared happy at the Rountree-Jablin house on Hearthglow Lane. What went on inside, however, was far from tranquil.

Kathryn Casey

Over the years, his office at the University of Richmond Campus became a refuge for Fred Jablin, as life at home became ever more chaotic.

Kathryn Casey

Piper's business card was as unusual as she was. On the immediate left, the dark winged angel of justice she painted for the card.

Piper's fish, Soul Rays. The children helped her paint it, and it memorialized the loss of children who grieved as her own children would just a few years later.

Kathryn Casey

In this glass block, Piper Rountree immortalized a young boy's loss: his father's suicide, a bullet suspended in the glass.

Kathryn Casey

At her clinic in an old house in Houston, Tina Rountree was well known, caring for many women in the neighborhood.

Kathryn Casey

Captain Stem had been with Henrico County for more than two decades, and he'd watched police work change with scientific advances.

Kathryn Casey

Henrico County Investigator Coby Kelley loved the challenge of an investigation, getting the suspect to open up and confess.

Kathryn Casey

Set back off the road, behind a strip center and a line of trees, the Homestead Suites was a perfect place to hide.

Kathryn Casey

Kevin O'Keefe, here inside Under the Volcano, at first believed he'd seen Piper Rountree on that Friday evening.

Kathryn Casey

Marty McVey seated in the chair in his reception area where Piper sat the day Coby Kelley followed her in the door.

Kathryn Casey

The Henrico County Courthouse, where the drama played out in February 2005.

Kathryn Casey

Henrico County Commonwealth's Attorney Wade Kizer believed that the Jablin case was among the most cold-blooded murders he'd ever investigated.

Kathryn Casey

To back him up, Wade Kizer brought prosecutors Owen Ashman (*left*) and Duncan Reid (*right*) onto the Jablin case.

Kathryn Casey

In Virginia, Murray Janus was the dean of defense attorneys, his list of clients including some of the most prominent names in the state.

Kathryn Casey

The trial would be Taylor Stone's first murder case, and it would prove to be a trial by fire.

Kathryn Casey

12

In Richmond, the Jablin murder investigation took on a set structure. Twice a day, the police officers involved—Hanna, Captain Stem, and Sergeant Russell—and the two prosecutors, Wade Kizer and the assistant he'd brought in on the case, Owen Ashman, met in the chief's conference room in Henrico P.D. headquarters. On the telephone, they patched in Coby Kelley in Houston, to discuss that day's objectives. They'd all worked together for many years, on cases that covered everything from robbery to fraud to murder. Yet that didn't mean the meetings were without tension.

At eight-thirty that Monday morning, Stem and the other police officers gathered wanted to arrest Piper Rountree. Her phone was used in Virginia to call her son, Paxton, who said that it was his mother's voice on the telephone. Ergo: Piper was in Henrico at the time of the murder, and she'd lied about it when asked point-blank by Kelley. Add to that the contentious divorce, especially the fight over child custody, and they had a motive for murder.

While all that was true, Kizer and Ashman didn't agree that the evidence was sufficient to hang a murder case on. "Having enough to make an arrest and enough to try a case are two different standards," Kizer told them. Before he'd consider pressing charges, the commonwealth's attorney still had the same demand: "We need to place not only Ms. Rountree's cell phone in Richmond but the woman herself;

otherwise a defense attorney will argue that Tina Rountree or someone else made that call."

Toward that end, on that Monday morning, November 1, Cindy Williamson turned over to investigators records she'd obtained under court order from Southwest Airlines and the Wells Fargo Bank. The debit card used to pay for the airline ticket was drawn off an account at Wells Fargo, an account taken out in the name of Jerry Walters. At that point no one involved in the investigation knew who Walters was or how he might be involved. That would be up to the Houston branch of the task force—Kelley and Dorton—to determine.

Meanwhile, in Henrico County, Stem told Hanna that he still wanted to know where the woman who'd flown as Tina Rountree stayed in Richmond. "She had to sleep somewhere," he stressed. "We need to know where. Wherever it was, someone there might be able to identify her." Despite not finding either Rountree sister on the hotel's registry, Hanna's gut was still acting up whenever he thought of the Tina Rountree registration at the Homestead Suites.

After that morning's meeting, for the third time since the murder, Hanna drove to the Homestead Suites. Again he talked to the desk clerk and the cleaning crew on duty, showing them photos of Rountree. Again no one could place the woman in the red dress in the hotel.

Disappointed, Hanna returned to the office and dug into the paperwork, helping Williamson draw up subpoenas for even more phone and bank records, asking for records from the days before and after the murder. As he went through the phone bills they'd already received from Sprint—Piper's cell phone company—Hanna noted that Piper Rountree appeared addicted to her cell phone, making dozens of calls every day. Combing through them all would be a formidable task, since the bills bore only unidentified phone numbers with no indication of the name of the person called.

While he worked on the cell phone bills, Williamson called Coby Kelley in Houston, giving him what information she could find on the Jerry Walters who'd opened the Wells Fargo Bank account. "I'll check into it," he assured her. But, first, he had something else to attend to.

That morning, Kelley, along with McDaniel, went to the Harris County District Attorney's Office in the criminal courthouse, a skyscraper on the rim of downtown Houston, to talk to one of the office's top prosecutors, Kelly Siegler, and ask for a search warrant for Piper Rountree's Kingwood house. They went over the evidence, showing Siegler what they had that tied Piper to the murder, principally the cell phone that revealed she'd lied about being in Richmond that weekend. Siegler frowned at the officers. Like the Virginia prosecutors, she wasn't sold.

"You need more evidence," she told them. "I can't take a request for a search warrant to a judge until you give me more to work with."

Violent thunderstorms hit Houston that Monday morning. Just after six, Charles Tooke called Piper at the house in Kingwood to make sure she was awake to meet with the investigators at Marty McVey's office at nine that morning. When she answered the telephone, Piper sounded upbeat, cheerful. "Everything's great," she said. Yet before Coby was ready to head to McVey's office, he heard from his office in Henrico. Piper had called, saying she wouldn't be meeting with him due to the bad weather.

Later that morning, after the rain stopped, McDaniel called McVey. "Is she coming?" he asked. "Is she going to talk to us today?"

"Piper talked to a civil attorney in Virginia about custody, and what she has to say could hurt her when she tries to get the kids. She's not going to talk to you," McVey said.

With that, Kelley took the telephone. "What's the deal? Are you representing her?" he asked the white-haired lawyer.

"No, I'm just a friend relaying information," McVey said.

With that, the investigators' day started on a disappointing note.

Although the Jablin case had been in the forefront all weekend, Breck McDaniel had other matters to attend to, including three murder trials to testify at that week. So he offered to stay in touch, to do what he could to help Kelley and Dorton, but he had other commitments.

The school bus stop where Fred Jablin had taken Callie each morning was quieter than usual that Monday morning. "There was a really eerie feeling about things. It seemed odd without Fred there. He was missed," says one of the neighbors who'd often talked with him. "We were all wondering who'd killed him. Was it Piper? If it wasn't, were the rest of us in danger?"

At the same time at the University of Richmond, Dean Ruscio and his staff were looking for a way to continue Fred Jablin's classes. They called in a part-time professor who'd filled in when professors had been out in the past, and he agreed to take over, but the materials he needed were in Fred's office, secured behind police locks and crime scene tape. To give him something to work with, the students were contacted and asked to bring in their work and the class syllabus, and the first session was held back in their old classroom. Many felt uneasy being in the room where they'd last seen their murdered professor. Some cried and one ran out of the room in tears, unable to continue. Soon afterward, Fred Jablin's classes were moved to a different classroom, one without so many painful memories.

Dean Lowry called Piper that Monday. Charles had alerted Dean, calling him to tell him of Fred's death and that Piper was worried she was a suspect. "She's trying to prove she was in Houston," Charles told him. "I tried to explain cell phone records to her, but I'm not sure she understood."

In the past, Piper had told Dean about Fred and her marriage, claiming that her ex-husband had always treated her not as a wife but as his student. She also charged that Fred was the one who'd had the affair, and that she'd known because he'd often been out late at night. Her affair with Dr. Gable, she said, was merely a way "to rub Fred's nose in it, like he'd rubbed mine."

On the telephone that morning, two days after the murder, Piper told Dean what she'd told Charles, that she wasn't anywhere near Richmond, Virginia, the previous Saturday. She'd been in Houston at the Volcano bar, where she'd met a man who'd walked her home. But when Dean asked questions, Piper said she couldn't recall if she'd left the bar at ten that evening or remained there until two the following morning, when the bar closed.

"Piper, I want you to go in your backyard and wait there for four hours and then call me back," Dean said, wanting her to understand what a long period of time she was talking about. "And I want you to consider what the police will think if you tell them that you can't say where you were for a full four hours."

As he listened to her, Dean had the unmistakable impression that Piper was lying. It bothered him to think she could have become desperate enough to commit murder, but as bitter as she'd been about the divorce, he thought he almost understood her desperation.

Without Piper asking, Dean proceeded to explain the way cell phone records worked, saying that they recorded what towers her phone hit off whenever it was in use, or when a call was made to the number or from the number.

After a pause, Piper said, "You know, I lost my cell phone. I've been using one of Tina's."

While Piper never confessed, Dean interpreted her long silences on the telephone as indications that the reality of her situation might be sinking in. Yet, it seemed obvious to him that she didn't understand what he was telling her, that with or without the telephone, the records would reflect when and precisely where it had been used during the previous weekend. As smart as she was, Piper couldn't seem to or appear to want to grasp why the cell phone records could be important. Trying to get through to her that this was something she needed to consider, Dean told her that the cell phone could either clear her, by showing she was in Texas at the time of the murder, or cast even more suspicion on her, if the phone had been used in Virginia.

With that, Piper asked a favor. She asked Dean if she could use his credit card to get a new cell phone. He refused, and the conversation became strained, with even more silence. Finally, he began to suspect that she realized he didn't believe she was in Texas the night Fred was murdered. Yet, he never asked if she'd killed Fred, and she didn't offer the information.

"I've heard that they serve warrants on people in the middle of the night, while they're sleeping," Lowry advised her. "Piper, you need to find a good attorney."

In Virginia, Wade Kizer turned over many of the day-to-day aspects of the case to the commonwealth's assistant attorney, Owen Ashman. A single mom with a long neck, wide-set eyes, pin-straight dark blond hair, and an all-business manner, Ashman came from an old Virginia family, one that traced its roots back to the Revolutionary War. Her first name, Owen, was a family name, her great-grandmother's maiden name, and she had followed her father, a commonwealth's attorney in another county, into law.

Although she had three other ongoing murder cases, the

Jablin case struck a special chord for Ashman, who'd lost her own father at the age of sixteen when, at just forty-three, he died of a sudden heart attack. She identified with all the pain and confusion the three Jablin children were undoubtedly going through, understanding how devastating the loss of a father could be. Like Kizer, Ashman had gone to the University of Richmond law school. She'd passed the bar in 1991, and worked in Virginia Beach for ten years, Newport News for seven, and had been in Henrico for five. Compared to Henrico, Newport News had been like the Wild West, with guns and drugs. "I'd interview a witness one week, and the next that same person would be dead, a murder victim," she says. "I got a lot of experience, quickly."

The Jablin case, based on the growing importance of Piper's cell phone and bank records, reminded her of a case she'd prosecuted years earlier, involving a husband who'd claimed his wife disappeared while he'd been at work. The man's cell phone records told another story, showing he was close to home that day. And his wife's bank records revealed that before her body was found, when she was still presumably alive, he'd used her credit card to buy his girlfriend jewelry.

That Monday, the first after the Jablin murder, Coby Kelley used two cell phones, alternating charging them, as he relayed information back and forth to Ashman in Richmond. The records were continuing to trickle in slowly under the court orders Ashman was writing with Hanna and Williamson. Finally, at about 1:10 P.M., something substantive arrived from the Wells Fargo Bank via the office fax machine: the purchases made on the Jerry Walters debit card during the days preceding and the murder.

On the bank-card records, Ashman and Williamson quickly zeroed in on one entry, a purchase from the Paris Boutique Wigs that sold on the Internet site wigsalon.com.

The entry that caught Williamson's attention was invoice 430211, for the amount of $261.73, including tax and delivery charges. When Williamson called the company's main office, in Miami Beach, Florida, she was told that the bill was for the purchase of two "Stevie" wigs, long and flowing past the shoulders in layers that the website touted as promoting a "sensuous essence." One was a brash red called Paprika Root, the other Frosti Blond, long layers of blond highlights. Both the wigs, as per the instructions that came with the order, were shipped to a Jerry Walters at an address on Kingwood Drive in Kingwood, Texas. When Owen Ashman looked at a map on the Internet, she discovered the address wasn't far from Piper Rountree's rented house.

When Ashman called the wig company and talked with Eleonor Ceballos, the clerk who'd taken the order, Ashman learned something else interesting. When the original order came in on the Internet, both wigs were on sale at a reduced price. By the time the wigs were to be shipped, the blond wigs that they had earmarked for the sale were sold out. The red wig was shipped without the blond one, accompanied by a note that said the blond wig would be shipped when it again became available at the sale price. The day the Paprika Root wig arrived, October 25, five days before the murder, a woman had called and talked with Ceballos, upset that she hadn't received the Frosti Blond wig.

"The customer told me to ship the blond Stevie wig at full price on a rush, overnighting it to the same address," Ceballos explained. The woman wanted the wig enough not only to pay full price, an additional $44, but Federal Express charges of $34.95 to have the wig by October 27, the day before someone traveling as Tina Rountree boarded the Southwest Airlines flight to Baltimore that connected to Norfolk, Virginia.

"She said that it was urgent she get the wig," said Ceballos. "She absolutely had to have it."

If Piper Rountree had committed the murder and used the wig as a disguise, then, Owen judged, the initial purchase a full nine days before the killing showed how long the murder had been planned. Another entry on the Jerry Walters debit card records backed up that logic. Also on the twenty-fifth, at 4:30 P.M., the same day the woman was putting a rush on a blond wig, someone had tried to use the debit card to purchase a ticket to Norfolk on the Southwest Airlines website. The purchase was declined for lack of funds. That situation changed days later when a thousand-dollar deposit went into the account. Then, on October 28, this time at the Southwest ticket counter at Houston's Hobby Airport, the card was used again, to purchase the round-trip tickets from Houston to Baltimore to Norfolk and back in the name of Tina Rountree.

What the records showed about the card from that date forward was just as interesting. Throughout the rest of the week, Thursday, Friday, and Saturday, the debit card was used in the Richmond area. The first time was on Thursday, October 28, shortly after the flight arrived, at a Shell gas station in the Norfolk area, near the airport.

Then, the bank records brought the card into the West End, within a small cohesive area, near the intersection of Cox and Broad Streets, in the same area as the Sprint records indicated Piper Rountree's cell phone was used at 4:36 on the morning of the murder.

First, it was used at 10:57 on October 29, the day before the murder, to withdraw $200 from an ATM at an East Coast gas station on Broad Street. Later that day the card was used at a 7-Eleven convenience store just down the street. Still later that same day, a purchase was made on the card at a nearby CVS pharmacy: $46.95 in merchandise.

Hanna and others fanned out across Richmond to follow the new leads, while Ashman added another assignment to Coby Kelley's growing Houston list: to go to the Kingwood address listed for Jerry Walters, the place where the wigs were shipped, and try to talk to Walters, to determine who he was and how he was connected to Piper Rountree.

In Richmond that afternoon another piece of evidence made its way to Wade Kizer's desk, the autopsy performed by Dr. Deborah Kay, a Henrico County medical examiner, on the body of Fredric M. Jablin. Kay described Jablin as a "well-developed, thin, white male, with his hands in paper bags." The bags had been put on by police at the scene to protect them until the M.E. had time to test for gunshot residue, to determine if Fred had fired a weapon. During the autopsy, Kay had administered an atomic absorption (AA) analysis, capable of finding even trace amounts, and found no indication Fred Jablin had fired a gun. Dr. Kay also found no defensive wounds on his hands, indicating there hadn't been a struggle with his assailant. The lack of gun residue on his clothing indicated he hadn't been shot at a close range.

A wound Kay labeled gunshot number two was fairly innocuous, entering the soft tissue of Fred Jablin's right arm and exiting without hitting any vital tissue, back to front, downward, from the right to the left. That bullet was recovered on the crime scene. The more serious wound was one Dr. Kay labeled gunshot number one, a bullet that entered Fred's body from the back, piercing his lower right side. Inside his body, the bullet did massive damage, ripping through his spleen, kidney, liver, diaphragm, and aorta. Gunshot wound number one was the fatal one.

When ballistics reports came in, the bullets were identified as .38 or .357 caliber Federal Hydrashock, nickel bullets, designed with a hollow point to inflict significant damage when entering a body. The most likely scenario, from the angle of the bullets, was that the killer stood near the garage, perhaps waiting for Fred as he came out to get his morning newspaper, then approached him and shot. That there were no casings found suggested that the murder weapon was probably a revolver, a type of weapon that didn't eject empty casings after firing.

"This type of bullet is designed to kill," Kay would later say.

What the doctor couldn't say was a subject that was being talked about all through Richmond: Would Fred Jablin have lived if he'd been found sooner? Dr. Kay wasn't sure. In her estimation, due to the extent of the wounds, he could have died almost instantly or he could have hung on for an undetermined period of time.

That Fred lay dead or dying in his driveway for more than an hour after the first 911 call infuriated many in Richmond, who believed the first officers on the scene should have found him, perhaps in time for paramedics to have saved his life. But those on the scene, including the McArdles, who'd found the body, didn't believe the officers had done anything wrong. Bob McArdle had watched Maggi and his partner comb the street with their flashlights, searching for the source of the shots. Even when Bob and his wife, Doreen, stood at the base of the Jablins' driveway talking to their neighbor, they couldn't see the body until after sunlight broke. "It was so dark out there, it was like looking for a needle in a haystack," says McArdle.

In his office, Kizer looked over the autopsy, while throughout Richmond the investigation went on.

That afternoon, Hanna drove to the CVS pharmacy where the Jerry Walters debit card had been used the day before the murder. When he showed the clerk who'd rung up the purchase the copy of the receipt, the man couldn't identify the pretty woman in the red dress in Hanna's photo as the person who'd made the purchase. But Hanna didn't walk away empty-handed. When he tracked down what items were purchased on the receipt, another piece of evidence clicked into place. That $46.95 charge, he discovered, was for the purchase of makeup, including mascara and lipstick, premoistened towelettes of the type used to remove makeup, and something that couldn't help but catch his eye: a box of latex gloves, size small.

Before calling it quits for the day, a disappointing one, in Houston, Kelley and Dorton drove back out to Kingwood, to the address on Kingwood Drive where the wigs had been shipped. What they found wasn't a house or an apartment belonging to a mystery man named Jerry Walters, but a Mail & More store, where packages were shipped and private mail boxes rented out to individuals.

By the time they held the second meeting of the day on the Rountree case, that evening, Kelley had one more piece of information for Kizer, Ashman, Stem, and the others in the Henrico conference room. Not only had Kelley and Dorton tracked down the address to the Mail & More store, they'd discovered that one of the postal boxes in the store was registered in two names: Jerry Walters and Piper Rountree.

"What more do you want?" Stem asked Kizer at the meeting the following morning. The two men were old friends and this was a familiar argument, the police wanting to make an arrest and the prosecutor holding them back, asking for more. Stem argued that the case had come together enough to arrest Piper Rountree. He wanted her house searched and her car impounded, before she destroyed evidence. Kizer, on the other hand, wasn't about to risk walking into a courtroom without enough evidence to get a conviction.

"Find someone who saw her in Richmond that Friday or Saturday. Someone who can positively identify her," he said again. "Then we'll get a warrant."

In Kizer's mind, it wasn't like they were dealing with a serial killer who would kill again. Piper Rountree was dangerous to one person—Fred Jablin. He was the only one she had a motive to kill, and he was already dead. "The police aren't the ones in the courtroom trying the case," says Kizer. "They don't understand what you need to get a conviction."

That morning, Tuesday, November 2, Coby drove Robin Dorton to the airport to catch a flight back to Richmond. Dorton had a family emergency and was needed at home. At the same time in Richmond, Chuck Hanna was going over cell phone records again with Cindy Williamson when something caught his eye, an entry for a phone number he recognized. "That's Papa John's pizza," he told Williamson. "I know, because I call it to order pizzas myself."

"Do you keep a database with your customer's names and records of purchases?" Hanna asked the manager at the pizzeria when he got him on the telephone.

"We do," the man said.

"Would you check to see if you have a Rountree?" Hanna asked.

Minutes passed while Hanna held onto the telephone. When the man got back on, he said, "Yes, we have a Rountree. She called from the Homestead Suites on Broad Street, and we delivered the pizza to Room 171, last Thursday evening."

"This could be it," Hanna said as he hung up the telephone. He'd had a feeling about that hotel ever since the first day of the investigation. Plus, it was strategically located, near the first phone call the morning of the murder and every place the debit card had been used.

That afternoon, Hanna and his partner, Bill "B.K." Kuecker, drove to the Papa John's with a photo of Piper Rountree, hoping to secure what Kizer said he needed, an eyewitness who could place Rountree in Virginia. Once there, they waited while the manager shuffled through a pile of receipts, finally coming up with the one for a large pizza delivered to a Rountree at the Homestead Suites. When the driver who'd made the delivery arrived, however, he couldn't help the officers with their problem. Before he even looked at the photo, he said, "I've probably delivered a hundred pizzas since this one. I couldn't identify anybody."

"Would you look at the photo for us," Hanna asked. "Just to be sure?"

The driver agreed. Hanna handed him the photo and the man was quiet, staring at it, thinking. "That doesn't remind me of anyone," he finally said. "Sorry, but I can't help."

"If someone stopped you in the parking lot and took the pizza, if you hadn't actually delivered it to the room, would you remember that?" Hanna asked.

"Yes," he said. "That I'd remember."

"Thanks," Hanna told him, and moments later he and his partner were back in their car headed toward the Homestead Suites. This time he had a room number: 171. But when the clerk at the front desk looked the room registration up, the name of the occupant that night didn't answer any questions, just brought up more possibilities. "Last Thursday night, that room was in the name of a Jerrilyn Smith," the woman said.

"Damn," Hanna said. "Who the heck is she?" Could she have been a friend of Piper Rountree, a hit woman, or someone totally unrelated to the case?

Back at the office, Hanna put the name Jerrilyn Smith through the computer and nothing of interest came up. Somehow, he knew this Smith woman was involved, but how?

For the rest of the day, Chuck Hanna made cold calls to the phone numbers on the cell phone records supplied by Sprint, the one for Piper Rountree's cell phone ending in 7878. One by one he wrote in who the numbers were assigned to. "Who are you?" he asked when he called.

"Who are you?" more often than not the people replied.

Meanwhile, that morning, Danny Jamison, the crime scene officer, met with Michael Jablin at the Hearthglow house.

Michael talked with the investigator while Mel and Barbara Boyd, the neighborhood mom who'd cared for the Jablin children after the murder, combed through the house with a list, looking for the clothing the children had asked for to wear to their father's funeral. The meeting was a somber one, Michael Jablin still looking stunned.

Afterward, with the bank records in hand, Jamison went to the places in Richmond where someone had used the Jerry Walters debit card, with subpoenas for any and all surveillance tapes from the day the card was used. Back at the office he watched them. On the tape from the 7-Eleven convenience store, he noticed a woman with long blond hair who looked enough like Piper to arouse his interest.

Jamison packed up all the tapes and sent them to the state crime lab for enhancement.

In Houston, Coby Kelley walked into Tina's Village Women's Clinic hoping for a talk. Instead she was out, but he talked with Melissa Hunt, a medical assistant who'd worked for Tina for years. "Did you see Tina last Saturday?" Kelley asked.

Hunt didn't hesitate. "Sure, she was here until about twelve-thirty, seeing patients. Then she got a phone call, and she left, in kind of a hurry."

"Is there any question in your mind that she was here working in the clinic last Saturday?"

"No," Hunt said. "No question at all."

Another piece of evidence had clicked into place: Tina Rountree was in Houston on Saturday, and Hunt would be able to testify to that, if need be. Tina, it was obvious, wasn't the woman in Richmond. Despite Homeland Security, it appeared someone else had flown to Virginia under her name.

That fact determined, Kelley concentrated on the next issue at hand: Jerry Walters.

With the bank records and those from the Mail & More,

Kelley easily tracked down Walters, who was working south of Houston in Victoria, Texas, on his cell phone. "I thought you were the vet," Walters growled when Kelley called. "My bloodhound, Bertha, is in the hospital. My relationship with that dog is the longest I've ever had with a female, except my mother. What's this about?"

Kelley was careful. He didn't know if Walters was involved with the murder, and he didn't want to reveal too much of his hand.

"I'd just like to talk with you," Kelley said. "I'm investigating the murder of Fred Jablin."

Reluctantly, Walters agreed to meet Kelley later that afternoon, but then he called hours before the meeting to say he was headed back home to Baton Rouge. "My dog's taken a turn for the worse," he said. "The vet's got her on oxygen. I have to go."

At first Walters balked when Kelley said he'd drive there to meet with him, but he finally agreed.

As soon as Kelley hung up the telephone with Walters, he called the Baton Rouge P.D. and made arrangements for an officer from that agency to accompany him to the meeting. Kelley didn't know Walters, didn't know if he was dangerous, and he wasn't taking any chances.

That afternoon, Kelley drove the four hours to Baton Rouge, hooked up with the officer assigned to accompany him, then met Walters at a Starbucks not far from his home. It was obvious from Walters's demeanor that the muscular man dressed like a cowboy wasn't happy to be involved. When Kelley asked when he last saw Piper, Walters recounted the previous week, when he'd driven into Houston and stayed at her home in Kingwood on Monday night.

"How did she seem?" he asked.

"She was fine, great, looked happy," he said. What he didn't say was that she certainly hadn't looked like a woman contemplating murder.

When asked, Walters detailed their relationship, saying they'd dated about a year after Piper arrived in Houston, then remained friends and just started going out again that fall.

"Where were you Friday and Saturday?" Kelley asked.

"Saturday I was at a football game in Baton Rouge," Walters said. "And there are lots of people there who saw me."

Kelley took down the information on Walters's alibi, and then turned his attention to the bank account. Did Walters know that his account had been used the previous week to purchase wigs, a ticket to Virginia, and at stores throughout Richmond during the days preceding and the day of the murder?

"Yes," Walters said. "I know."

With that, Walters explained that he'd opened the account in August 2004 at a Wells Fargo Bank inside a Randall's grocery store in Kingwood, not far from Piper's house, for her to use during the time she was declaring bankruptcy. It was a savings account with a debit card, one that allowed her to make deposits and withdraw money to pay for purchases. Later, he'd discovered, she'd also had checks issued on the account.

He'd first heard about Fred's murder the night after the killing, when Piper called him and said simply, "Fred is dead."

She'd asked Walters to come to Houston, to stay at her house. Feeling uncertain about all that was happening, he didn't go. As they'd talked, something she said caught Walters's attention: that she'd lost the debit card to the Wells Fargo Bank account.

"The next morning, I got on the Wells Fargo website, and found out that the account was overdrawn and that it had been used in Virginia," Walters said, frowning. "So I called her back. She said that the card had been stolen at the health club, while she was playing tennis. I asked her how somebody could use it without the access code, and she said she'd written it on the card, backward. Then she asked me not to report the card stolen to the bank. But I did, and I closed the account."

That was information Kelley knew he could check on after he'd looked into Walters's alibi. What he asked about next was the P.O. box, number 162, with Walters's name on it. "Piper had that box," Jerry said in his thick, whiskey voice. "She asked me to pick up the mail for her there, so she added my name to it."

That was information Kelley had already received from the manager at Mail & More. At least that part of Walters's story he knew to be true.

Kelley's third set of questions for Walters involved telephone calls made from Piper to his home the day of the murder. Had he talked with her?

Although one of the calls, according to the records, had lasted eleven minutes, Walters said he was sure he hadn't talked with her early that day. The tension built as Kelley asked repeatedly if Walters could be wrong, if he might have talked with Piper that day. If so, Kelley wanted to know if she'd told him where she was and even what she'd done.

"I'm telling you that I may have talked with her. I talk with her all the time. But if I did, I don't remember it," Walters growled angrily. Once he calmed down, however, he remembered that he did get a voice mail from her that day, during a time when Kelley knew Piper Rountree's Sprint records showed her cell phone was bouncing off towers in Virginia.

"Are you sure she was the one on the telephone message, not someone else, like her sister, Tina?" Kelley asked. "Maybe someone who sounded like her?"

"It was Piper's voice," said Walters, punching another hole in Piper Rountree's story that she hadn't been in Virginia that weekend.

Then Walters told Kelley something else: that earlier that

same week, the Monday or so after the murder, Piper had called him again, asking him to stay at her house, in case anyone showed up looking for her. She said she and Tina wouldn't be there, that they'd checked into the Houstonian Hotel. As he had earlier when she'd asked him to go to Houston the night of the murder, Walters refused.

13

Wednesday morning, November 3, the fifth day after the murder, Investigator Chuck Hanna sat in the morning meeting with Stem, Russell, Kizer, and Ashman, and felt frustrated. In his opinion, the Jablin investigation had stalled. Piper Rountree, lawyer or not, had left a bunch of clues for them to follow, but none of them were leading to what Kizer said he needed—an eyewitness willing to positively testify that she was in Richmond the day preceding or the day of the murder. At the meeting, the officers and prosecutors snapped at each other, Hanna figured out of the same disappointment he felt. Afterward, Hanna got on the telephone with Kelley, filling him in on the tension in the meeting.

"Man, you owe me," Hanna said, and both investigators laughed.

That day, at Hanna's request, Steve Byrum called Piper Rountree on her cell phone that ended in 7878, the same one police were subpoenaing records on and the same telephone that she'd told Dean Lowry she'd lost. After answering, Piper immediately gave Byrum another phone number to call and hung up. When Steve dialed the new number, it rang the Houstonian Hotel. He asked for Piper Rountree and was connected to Room 220. Without Piper knowing, Hanna recorded the conversation.

For days, Piper had left messages on Byrum's phone: "I need to talk with you." Now that he'd called, Piper was friendly, but she didn't seem overly nervous or concerned. "I just wanted to see what was going on," she said.

"You said you needed my help," Byrum said to her. "What is it?"

"I need to know how my kids are and where they are," Piper said.

"You know they're not here, with me," he said. If the police had the children, Steve advised her, she needed to get a lawyer.

"I'm working on that," she said. "Have the police been to see you?"

"Yes," he said. "I don't know why. I guess your children told them we had some kind of a relationship or a close friendship."

Then she told Byrum she was staying at the Houstonian because the police kept bothering her. She gave him a new cell phone number, but said it could change again. Byrum asked if she'd been in Richmond on the day of the murder, and Piper said, "No."

"You need to be going and taking custody of your children," Byrum said.

In Virginia, Chuck Hanna called Southwest Airlines to track down anyone who interacted with the person who traveled as Tina out of the Norfolk airport. He grew hopeful when he discovered that the original airline ticket had a scheduled return for Sunday, not Saturday. When the woman arrived at the airport on Saturday morning at 8:29, two hours after the murder, "Tina" had the ticket changed by a clerk at the Southwest counter for the next available flight to Houston. Since she'd changed a ticket and not merely checked in for the flight, Hanna figured the clerk had spent more time with the woman, improving the chance that "Tina" might be remembered by the ticket agent. When Hanna tracked the clerk down, she quickly agreed to meet with him.

Minutes later Hanna and his partner were in his car and driving to Virginia Beach, a two-hour trip, to the woman's house, hoping for the break that could ease his frustration and result in an arrest warrant for Piper Rountree. When they arrived, they stood outside the apartment on a chilly fall morning, while the ticket agent inspected a copy of the ticket.

"Do you remember changing this ticket last Saturday morning?" Hanna asked.

"Not a thing about it," she said with a frown. "To tell the truth, I've been working double shifts. That was my last day in a long string of days, and I'd changed a lot of tickets for a lot of people. But if you want to show me the photo, I'll look at it."

Hanna took the photo of Piper Rountree in her red dress from his file and handed it to her.

"No, I don't," she said after sizing it up. "Not at all."

Later that day, Dean Lowry talked with Piper on the telephone and gave Piper similar advice to what Steve Byrum had offered earlier that morning: that it was time for her to stop acting guilty and scared and start acting like a concerned parent.

As with their previous conversation, there was hesitancy in her voice. Lowry thought, *Piper did it, and she knows that I think she did it.*

When Dean asked Piper if she still had her cell phone, the one she'd answered that morning when Steve called, she said, "No, I threw it out, in pieces."

As she said those words, Dean's mind flashed an image: Piper driving down a highway, scattering pieces of her cell phone out the window of her black Jeep. Perhaps she believed the record of the calls was on a chip inside the telephone itself, not recorded in the business office of the cell phone company? he thought. Maybe without the telephone, she assumed the police wouldn't be able to trace where it was used?

In Richmond, at two that afternoon, a funeral service was held for Fred Jablin at Congregation Beth Ahabah. Despite Fred's stature at the university and in the academic world, on the advice of police, there'd been no obituary in the Richmond newspaper. Still, two hundred people showed up to honor a man and a scholar. When Professor Ciulla pulled up and parked the car, she heard on the radio that her candidate in the presidential election, Senator John Kerry, had just conceded. It added to the gloom she felt and seemed somehow fitting. Inside the synagogue were others from the University of Richmond waiting for the funeral to begin, many of them students who'd availed themselves of a bus the school had supplied.

Many of those attending looked for Piper in the crowd, but she was noticeably absent. While most mothers would rush without hesitation to their children's aid, a full five days after her ex-husband's murder, Piper Rountree remained in Houston. She'd cried throughout her interview with Kelley, claiming she was worried about her children's welfare, yet she'd still made no attempt to be with them. Instead, Piper was hiding out in a hotel room, trying to avoid the police.

Michael Jablin and his wife, Elizabeth, along with their two children, brought Jocelyn, Paxton, and Callie to the service. Many cried as the Jablin children entered the synagogue. Michael had his arm around Paxton, who looked traumatized, while Callie and Jocelyn were pale and silent. Melody Foster and others from Kingsley were there, to say goodbye to a friend and a neighbor. With them were their children, the Jablin children's friends. The parents were having a difficult time explaining such a tragic turn of events to their children. Dads weren't supposed to be murdered in their driveways in quiet, upper-middle-class neighborhoods.

During the service, Rabbi Beifield talked of Fred, his up-beat attitude and his devotion to his children, mentioning that Fred had wanted them to have the strength of their religion, after the divorce enrolling them in religious programs, including Jocelyn in the youth group.

In the front row of the synagogue Paxton's friends sat wearing yarmulkas, to support him. While Elizabeth Jablin cried inconsolably, Ana, the children's beloved nanny, gently rubbed Jocelyn's shoulder and whispered to comfort her. At the end of the service many dried tears as Michael, Elizabeth, and their family, which now included Fred's three children, stood to leave. There would be no burial in Richmond. Fred's body was to be interred in New York, near his parents.

Afterward, a reception was held at the Boyds' home. PTA moms supplied ham biscuits and brownies, while Paxton's friends waited with excitement for his arrival. That afternoon they presented him with a gift, a brand new skateboard, something for Paxton to remember them by.

In Houston, Coby Kelley had a break: the name of the Houston Southwest Airlines ticket agent who'd checked Piper in the morning of her flight. He'd already heard from Hanna that that approach had been fruitless in Norfolk, but Kelley hoped he'd have better luck at Hobby Airport with a woman named Kathy Molley.

When he arrived, Kelley found Molley to be a friendly woman with curly blond hair and dark eyes.

"Would you look at this ticket and see if you remember anything?" Kelley asked.

She agreed, and he handed her the ticket.

"Yeah, I remember this," she said. "She was a really cute woman, nicely dressed. I'm not a lesbian, but she was really attractive. And she was wearing a blond wig."

As Molley went on, she described the encounter, saying she remembered because she thought the woman's name,

Tina Rountree, was "cute." But there was something else.

"You know," Molley said. "This woman checked a gun."

Kelley tried not to show his excitement. "Would you take a look at the photo we have and see if you remember this woman?" he asked.

Molley sized up the photo of Piper and didn't hesitate.

"That's her," she said. "That's the woman who checked the gun."

As Kelley listened, Molley explained in greater detail what had happened that morning. The woman, she said, was in a rush, wanting to get on the next available flight to Norfolk. When she'd checked her in, the woman said, "I have a gun in my luggage that I need to declare."

Despite being forthcoming about having the gun, the woman hesitated when Molley said it had to be removed from the suitcase to be inspected. "I explained that I needed to see it, and she kept fumbling around in her suitcase, trying to get it out. I thought maybe she was hoping I'd forget about it," Molley said. "I told her if she was traveling with it, I needed to see it."

Finally, the woman Molley had just identified as Piper took the gun out of her luggage. It was in a small beige case, perhaps a foot long, encircled with a bicycle-type combination lock. At that point Molley asked her to open the case, take out the gun and check it to make sure it was unloaded. Piper did as asked and showed Molley the gun, secured by a second cable lock that went through the trigger.

Once she was sure the gun wasn't loaded, Molley gave Piper the required forms to fill out and called over one of the Transportation Security Administration personnel at the airport to inspect the weapon.

The TSA agent who responded that day was Allan Benestante, a tall, thin man with a furrowed brow and dark eyes hidden in shadows, who worked as a manager screening luggage.

Kelley's luck was improving.

When Molley inquired, she discovered that Benestante was at the airport working. A few minutes later the luggage screener arrived at Molley's station and finished the story of the woman with the gun for Kelley.

While it wasn't unusual for people to transport guns in checked luggage, Benestante said it was unusual for women to do so. "She told me it was her father's gun, that she was taking it to him," he said.

The woman was white and middle-aged, Benestante said, and the gun was a chrome revolver with a short barrel, a .32 or .38 caliber, with a wood or composite grip. Without saying anything, Kelley smiled, realizing that Benestante's description matched the ammunition expert's assessment of the type of gun that fired the fatal bullet.

Inside Piper's gun case, Benestante said he'd also noticed a small box of ammunition. While he inspected the gun, the TSA agent said Piper had looked nervous, fidgeting. He'd thought little of it, however. Since 9/11, travel was stressful and many people looked frazzled. After he checked around the gun for explosives and found none, Benestante placed the declaration Piper had filled out in the gun case and closed it, then waited while she locked it. The woman then repacked the gun case inside in her luggage. As she did, Molley remembered noticing a particular pair of shoes in the suitcase.

"They were really cute," Molley said. "And I told the woman that."

Then Piper's luggage, with the gun inside, was put on the conveyor belt and checked onto the belly of the airplane.

Everything had gone so well, Kelley was surprised at the next turn of events.

When he handed Benestante the photo of Piper Rountree, the security agent wasn't sure she was the woman he'd seen with the gun. "I think that's her, but I'm not positive," he said. "The hair is different and she looks thinner."

As soon as he finished, Kelley put in a call to Virginia, telling his sergeant what he had. "Do we have enough now?" he asked. "Can we pick her up?"

In Virginia, Hanna and his partner, Kuecker, were headed back to the Homestead Suites Hotel for another go around. They'd called ahead, and the manager on duty the night Jerrilyn Smith had checked in was working. They were eager to talk with her.

At first, Tomiko James didn't recognize the name of the woman they were asking about.

"Have you got a picture?" she asked.

"Yeah," Hanna said, pulling the photo of Piper out of his file and handing it to her.

James, a tall, friendly woman with skin the color of coffee with cream, smiled. "That's her," she said. "That's the lady who checked in as Jerrilyn Smith."

"Can you describe her, how she looked that night?"

"Well, she was thin, a white female, and she had a black hat and long blond hair," she said. "It wasn't cold out, but she was wearing a hat, coat, a scarf, and sunglasses."

As James recounted it, Piper had checked in at eight-fifty that Thursday evening, first as Tina Rountree, the name the reservation was under, for Room 171, a nonsmoking room, at a cost of $65.99 a night. The woman, who appeared nervous and paced the lobby, had identified herself as a leisure guest, not in Richmond on business. When she checked in, Piper had handed James a Texas driver's license in the name of Tina Rountree for identification. James wrote up the paperwork and returned the driver's license to her. Then Piper handed James cash to cover the bill and asked, "Can I change my name on the register?"

James wasn't surprised; people did that for a variety of reasons, including women trying to hide their identities while they fled abusive spouses. "Sure," James said. "What name do you want to be listed under?"

"Jerrilyn Smith," Piper said.

James made the change in the computer, and the woman left to go to her room.

The next morning, Friday, the maid found the room empty, and it was assumed Ms. Rountree/Smith had checked out, but that afternoon Piper returned, again paying cash and checking in as Jerrilyn Smith. Early the next morning she phoned the desk and checked out for good.

As soon as they left the Homestead Suites, Hanna called headquarters. "We've got a positive ID at the Homestead Suites," he said. "The woman identified the photo of Piper Rountree as a hotel guest for Thursday and Friday nights."

"Good," Sergeant Russell said. "I'll let the captain know."

In Richmond the case seemed to be falling in place, but unbeknownst to the investigators, in Houston events were taking place that could have stopped the momentum the Jablin task force had worked so hard to build.

That night a group of good friends met at their usual haunt, Under the Volcano, a popular bar across the street from a bookstore called Murder by the Book, and a block from Tina's clinic and Marty McVey's office. It was the same bar Piper had told Charles Tooke and Dean Lowry she'd been in the Friday night before the murder, where she'd met the man who'd walked her home.

Hidden behind palm trees and thick vegetation, the Volcano was decorated like an island beach bar, in bright Caribbean colors, the walls covered with folk masks and statues commemorating Mexico's Dia de los Muertos, the Day of the Dead, the annual fiesta during which it was believed the dead dropped in on living family and friends. The name of the bar came from Malcolm Lowry's 1947 novel, which told the story of the last day in the life of Geoffrey Firmin, a British consul and chronic alcoholic.

At the bar, the gang of friends who called themselves "Team Martini" had gathered for a bit of after-work revelry. They were an eclectic clique, mostly middle-aged professional men who'd met years earlier at the Volcano and forged a common bond. Over the years, they'd become close friends.

On this night, the members of Team Martini in attendance noticed Piper and Tina walk into the bar. Tina paced, looking markedly agitated, while Piper approached Cheryl Crider, the pretty blond bartender.

"Do you remember me? I was in here last Friday night," she said.

After sizing Piper up, Cheryl said she thought she did remember seeing Piper. But had it been Friday night?

"Something's happened, and I need to find the man I was talking to at the bar," she said. "It's really important."

Cheryl turned to members of Team Martini, who were laughing and talking, and asked if they remembered Piper. At first all of them said no. Then Kevin O'Keefe, a sandy-haired man with a degree in philosophy who made his living working on large generators, walked in the door to join the group. When Piper asked Kevin if he remembered seeing her in the bar the previous Friday evening, O'Keefe looked at her and said, "I think I do remember you."

"I met this guy and we had a couple of drinks. I was having wine. I don't remember what he was drinking," an excited Piper said as Kevin sipped his first Paulaner Munchen beer of the evening. This man she'd met, Piper told O'Keefe and Crider, had offered to give her a ride home, but she'd refused. Instead, he'd walked her to her sister's house, a few blocks away.

As he thought back, Kevin could picture Piper in the bar

the past weekend, among a sparse crowd of patrons, first standing around a thick wooden post in the center of the bar, then sitting on a bar stool talking to a man. O'Keefe, dressed in his customary T-shirt and shorts, didn't remember seeing Piper before that night, and he didn't recognize the man she was with.

"Why is this so important?" Cheryl asked.

"I got a call from the Virginia police," Piper said. "My ex-boyfriend I lived with four years ago was stabbed to death on Saturday morning, and they want to know where I was."

"Four years ago," O'Keefe said, considering that this sounded serious. "Well, where were you on Friday during the day?"

"At work at the Galveston courthouse," Piper said.

When O'Keefe asked if someone there could verify her presence in Texas, Piper said no one could. When he asked if she had any receipts that placed her in Houston that Friday or Saturday, she said she didn't. But if she could prove she was in Houston on Friday night at the bar, she theorized, certainly the police would realize she wasn't in Virginia before daybreak on Saturday morning. O'Keefe felt only confusion when Piper then mused, "But what if I had a really fast airplane?"

O'Keefe, who'd had a friend murdered in the mid-seventies, thought the conversation was becoming increasingly bizarre, but it was about to become even stranger. At that point Tina walked in and put her hand on Piper's shoulder.

"Piper, we need to talk outside, right now," she said. "It's about my period."

Once the women were outside, the men on Team Martini shook their heads, amazed at the two sisters and Piper's peculiar tale.

Minutes later Piper walked back in the bar and approached O'Keefe and Crider for a second time, asking for the best way to reach them, in case she needed them for an alibi.

Happy to help, they wrote their cell phone numbers on a sheet of paper and gave it to her. Piper thanked them and left. The bar patrons thought the night's strange interlude had ended and returned to their drinks and conversation. But half an hour later Piper Rountree walked through the Volcano's door again, this time leading an entourage that included not only Tina but two men in business suits, one with a ledger and legal forms.

"I need to have you two give me signed statements these men can notarize and give to the police," she said to O'Keefe and Crider. "I need to prove I was here on Friday night."

They looked at each other, and both started shaking their heads.

"You have our phone numbers, that's enough," O'Keefe said. "When the police call, you give them the numbers, and we'll talk to them."

Piper was insistent, beginning to argue with them, when Tina nudged her.

"Piper, let's go," she said.

Piper looked at O'Keefe and Crider, then at her sister. Saying nothing more, she turned and left, leading the odd procession from the bar.

"Now, that was really peculiar," O'Keefe said to the others.

"On Friday night I was at a bar called the Volcano on Bissonnett," Piper said to Coby Kelley on the telephone at eight-nineteen the next morning, Thursday, six days after the murder. Kelley was in his Houston hotel room at the time and had just finished his morning conference call to Richmond. "Two people can verify I was there," she said, then rattled off names and phone numbers for Crider and O'Keefe. It was the first communication Kelley had received from Piper since the morning she'd cancelled their meeting at McVey's office, and he wasn't able to record it. So he took notes, and then, while he was talking with her, picked up a tape recorder and repeated what she'd told him into it, while he had her on the telephone. Later he transcribed the notes onto a computer in Breck McDaniel's office at HPD.

Piper Rountree ~ Call where she provides alibi info, he titled the page.

In her account, Piper said she'd been at the bar from 8:30 until around 10:30 P.M. And she had witnesses, O'Keefe and the bartender, Crider, who would tell Kelley they remembered seeing her. There was another possible witness, she said, a man she connected with that evening who walked her to Tina's, a man she sometimes called Jerry and other times Steve. She hadn't been able to find him yet but was still looking.

"Which one is it, Jerry or Steve?" Coby asked.

"He switched back and forth," Piper said. "Sometimes he said his name was Jerry and other times he said it was Steve."

She had other evidence of her innocence as well. Her friend, Charles Tooke, would be able to show Coby an e-mail she'd sent to him on Saturday afternoon from her home computer. And she'd also called Charles from her house that afternoon. Finally, she'd remembered that one of her neighbors had stopped over with her daughter on Saturday afternoon about two-thirty selling Girl Scout cookies, and Piper had opened the door and talked with her. If that were all true, how could she have been on an airplane on Saturday that didn't land in Houston until four-forty?

The one bit of evidence Piper admitted that worked against her was that she still said she had talked with Paxton and Callie on that Friday afternoon. "Yes," Piper said. "That was me." Cell records showed that both calls had originated in Richmond, not Houston.

Always, her attention was drawn back to the children, and

when Piper pressed Coby, demanding to know if he would testify on her behalf and help her get custody, he answered, "It's out of my hands . . .

"The bottom line is that at 6:30 A.M. on the East Coast your ex-husband was murdered," he told her. "If you can verify where you were . . . that would be helpful. If you were with somebody, anybody, even a one-night stand."

When Coby hung up the telephone, he knew he had a lot more legwork to do before the Jablin case resulted in an arrest.

In Richmond at Henrico P.D. headquarters, an imaginary line had formed down the middle of the chief's conference room. On one side, Wade Kizer argued against arresting Piper, still wanting more evidence. Captain Stem, on the other side, couldn't believe Kizer was still holding out. He argued his men had more than enough evidence to make an arrest. Uncomfortably in the middle stood Owen Ashman, who was working with the D.A. in Houston, attempting to get a search warrant for Piper's Kingwood house. She understood both Kizer's and Stem's positions. When Kizer walked out of the chief's conference room and Ashman was alone with Stem and his men, she'd later say, "I got an earful."

Now they were all facing a glitch that could send the investigation back to the beginning. They had what Kizer said he needed, not just one person who identified Piper as having been the woman who traveled to Richmond as Tina, but two: the hotel manager Tomiko James and the Southwest Airlines ticket agent Kathy Molley. Yet Piper, it appeared, had at least two people and perhaps three or four who could place her in Houston late on the night before the murder, too late to make it to Richmond by Saturday morning. If that were true, they had a problem. Who traveled on the Tina Rountree ticket?

"We need more evidence," Kizer said.

Frustrated, Stem agreed that his men would keep digging.

Help arrived with the videos that showed up at Henrico P.D. headquarters that Thursday. After the enhancement done at the state lab, Danny Jamison could see a woman who looked like Piper walking into the 7-Eleven wearing what appeared to be a blond wig, a three-quarter-length dark coat, and gray sweatpants, carrying a shoulder bag. He noted that the wig looked similar to the Frosti Blond one purchased on the wigsalon.com website. Still, Kizer said, he needed more.

That afternoon, Henrico also received the final cell phone records from Sprint, fulfilling the subpoena. There were stacks of pages and hundreds of phone calls to dissect. On the day of the murder, Piper had begun calling early and hadn't stopped. The majority of the calls were to Tina, four times just between 7:19 and 8:40 A.M., and other calls that continued throughout the day. When Owen Ashman examined the records, she could only imagine what Piper might have been telling her sister, including, "I finally did it. I killed Fred."

Other calls had been to Piper's other sister, Jean, and to Jerry Walters.

In Houston, Kelley wasn't sure what to make of Walters. That Walters had been forthcoming with information made Kelley think he probably wasn't involved. He'd checked Walters's alibi, and it appeared he had been at the L.S.U. football game in Baton Rouge that weekend, as he'd said. Plus, Williamson had checked, and Walters's cell phone was used in Louisiana all that weekend, including when Piper called him at 8:44 on the morning of the murder. But he was still troubled by the phone call that the records showed had come in from Piper and lasted eleven minutes. "I wanted to know what she told him on the telephone that morning, if she'd confessed," said Kelley. Yet whenever he asked, Walters insisted that he didn't even remember talking with her.

* * *

At the University of Richmond, Vernon Miller had the unhappy task of cleaning out Fred Jablin's office. The police
had finished with it, and the university needed someone
willing to sift through the contents. Miller found the room,
with its bookcases and windows overlooking the campus,
cluttered but organized, with papers stacked in piles by
topic. Throughout the office, on the desk, walls, and shelves,
were photos of Fred's three children and their artwork,
proudly displayed.

Miller felt stunned by the loss of his former professor.
Fred had been a good friend and a mentor, and in Fred's office, Miller sensed his personality once again, for the final
time. In the clutter he'd left behind, Fred had left an indication of the man he'd been in life, one devoted to his children
and his work.

On his desk, Miller found Fred's notes on his latest work,
his investigation of courage, an irony that didn't escape
Miller as he stacked some material to be sent to Linda Putnam, Fred's collaborator on the books, some to Michael
Jablin for the children, and some to be retained at UR for a
memorial. "Maybe Fred had been able to draw courage from
reading the work of others," Miller pondered later. "It was
certainly the most difficult time of his life."

Meanwhile, at Henrico P.D. headquarters, Chuck Hanna
found something interesting while going through the phone
records. Earlier, he had called all the usual car rental agen-
cies at the Norfolk airport, and none had a record of a Tina or
a Piper Rountree renting a car on Thursday, October 28. But
as he worked his way through the phone bills, cold-calling
numbers that Thursday, he dialed a number and the person
who answered the telephone said, "Eagle car rental."

"I found it," he said to Williamson. "She called a car
rental agency."

"Did you ask them if they rented her a car?"

Hanna laughed, embarrassed. He'd been so excited when the woman answered that he'd just hung up the telephone. Minutes later he had Eagle back on the line. "Did you rent a car to a Tina Rountree last Thursday?" he asked.

"Yes, we did," the woman said.

Yes, Hanna thought. Finally.

"How could you answer that so quickly?" he asked.

"She's white, and we hardly ever have Caucasians coming in to rent cars," she said.

"Is the vehicle there now?" Hanna asked. When the woman said it was, Hanna told her to keep it there. "We're coming down."

Kelley's partner, Robin Dorton, was the one who drove to Norfolk to talk to the people at Eagle that afternoon. When he arrived at the small auto sales and rental agency on North Military Highway, Tarra Watford, a thirty-four-year-old clerk with her hair in braids, told him of her interaction with the woman who called herself Tina Rountree.

The woman had called earlier in the week, she said, asking if she could rent a car without a credit card. When she learned that she could, with her driver's license and a utility bill to show proof of her address, she wanted to reserve a compact car, one that Eagle rented for a mere fifteen dollars a day. Watford explained that Eagle, a discount agency, didn't take reservations but rented the cars on a first-come, first-served basis. The woman who identified herself as Tina Rountree argued the point, but Watford didn't bend.

Then on Thursday, October 28, just after 5:00 P.M., the same woman called from the Norfolk airport, to ask them to hold a car. At that time, all Watford had available was a maroon 1998 Ford Windstar van with a gray interior that cost $55 a day. The woman sounded miffed but said, "Okay, I'll take that."

Fifteen minutes later Watford looked out the window and

saw a cab pull up in front of the window. A petite, middle-aged woman got out with her luggage and came inside. She was in a hurry but polite, handing Watford a driver's license and a Time Warner cable television bill with the name "Tina Rountree."

At first, "Tina" said she wanted to return the car on Sunday—the day "Tina" had booked her return flight. Watford explained that the agency would be closed that day.

"I'll have it back Saturday, then," the woman said.

When Watford asked for a Richmond address, the woman gave her the address and a phone number for the Towne-Place Suites, the hotel Piper had stayed in over the years when she'd come to Richmond to see the children. "I normally stay with friends when I'm in Richmond," the woman had said. "But they're gone so I'm staying in a hotel. I'm thinking about buying a place there."

Tarra went over the agency's policies on gas and insurance, and then went outside with the woman to check the van's fluids. Twenty minutes after she arrived, the woman put down a $230 deposit and left in the rented van.

"You know, I usually take people's fingerprints when they rent cars," Watford said to Dorton. He cringed when she added, "But the pad was dried up, and I didn't have any ink."

On Saturday, the woman returned. She'd driven 342 miles, and Ray Seward, the owner, checked her in, gave her a $65 refund from her deposit, and then his son drove her to the airport.

Dorton pulled out the photo of Piper he had with him, one of her sitting on a rock near a lake. He also took out Tina's driver's license. Watford eyed them both, then pointed at the driver's license. "I think that's her," she said, identifying Tina as the woman.

With that, Dorton went outside to talk to Seward, a balding man. He verified what Watford had told Dorton.

"Could you identify the woman?" the investigator asked.

"Yes," Seward said.

Dorton showed him Tina Rountree's driver's license photo.

"That doesn't look like the lady I saw," Seward said.

Then Dorton pulled out the photo of Piper he had with him, one of her sitting on the rock.

"That's the lady," he said.

"How sure are you?"

"One hundred percent," he said. "She was a nice lady, attractive."

Seward then gave Dorton some bad news. The van had not only been cleaned after the woman in the photo rented it but it had been rented by another customer.

Still, Dorton hoped, perhaps there would be something left behind. When the crime scene unit arrived they did find evidence to collect: fingerprints, a fruit punch bottle, water bottle, candy wrappers, a bottle of lotion, a pair of black sunglasses, and something Dorton found very interesting— a key to a room at the Homestead Suites.

In Houston that afternoon, Kevin O'Keefe was at the Volcano bar, where he'd arranged to meet with Coby Kelley, who took notes as the Team Martini member recounted the events of the evening before. When O'Keefe got to the part where Piper referred to the murder victim as her ex-boyfriend, Kelley was surprised.

"That's weird," he said. "It was her ex-husband. They were married for eighteen years and had three kids. Are you sure she said boyfriend?"

"Yeah, boyfriend," O'Keefe said.

"Are you sure it was Friday night you saw her? Could it have been another night?"

O'Keefe thought that over. He knew he'd seen the woman, but could it have been another night? He'd been working a lot lately on a big project and often didn't get to hook up with his buddies at the bar as he normally did. He wasn't sure when he'd been there the previous week.

"I guess it could be another night, but it was sometime last weekend," he said. "I'll have to check my records."

Later on his cell, O'Keefe called Cheryl, the bartender. "Is it possible we saw that woman another night?" he asked. "Is there a way to tell when I was in the bar last week?"

"Yeah, we keep the bar tabs with your name on them," she said. "We can check. But I didn't work Saturday night, just Friday, and I saw her."

"Well maybe she's right and it was Friday night then," he said.

14

. . . .

Early Friday morning a call came in to Henrico P.D. headquarters, which was later recounted to Coby Kelley. The caller, a woman named Patti, said her friend, Carol Freed, a comic who worked at a defensive driving school, knew something about the Rountree case, including the possible involvement of Tina Rountree. McDaniel and Kelley drove to the school, in a small, tired-looking beige brick office building two miles from the Volcano, Tina's clinic, and all the other sites they'd visited while investigating the case. They walked through an overgrown courtyard, up a flight of stairs, and into the driving school, where clocks on the wall marked time around the world. The place was empty except for a man sitting behind a reception desk. McDaniel asked for Carol Freed, and the man said she wasn't in. Explaining he was an investigator with the Houston P.D., McDaniel handed the man his card and asked him to call Freed. The receptionist picked up the telephone and did as instructed.

"There are some police officers here to see you," he said, handing McDaniel the telephone.

"Ms. Freed, we're investigating the murder of Fred Jablin, and we'd like to talk with you," he said.

McDaniel heard a click and the line went dead.

With that, McDaniel returned his attention to the man behind the desk. "Tell Ms. Freed we're not going away," he said. "Tell her she needs to give us a call."

* * *

In Virginia, Investigators Thomas Holsinger and Stokes Mc-
Cune followed a lead off the bank records to a Miller Mart
convenience store in Williamsburg, Virginia, where the Jerry
Walters debit card was used just ninety minutes after the
murder. The two officers walked in and showed the photo of
Piper Rountree in the red dress to Tina Landrum, the assistant manager.

"Oh, yeah, I remember her," Landrum said, explaining
why the woman had made an impression. The woman in the
photo, she said, came in that morning and hung around "the
junk aisle," where the silk flowers and novelties were displayed. Landrum noticed the woman walk to the beer cooler
and then the ATM. The ATM rejected her card. The woman
then approached Landrum, behind the counter, and asked if
she knew what the limit was on the ATM. "I told her it was
whatever she'd arranged with the bank," Landrum said.

As they talked, Landrum smelled alcohol on the woman's
breath. The next thing Landrum knew, the woman was gone.

There was one other thing Landrum remembered: The
woman in the photo was wearing a longish blond wig.

McCune asked if she were sure that the woman in the convenience store that day and the woman in the photo were the
same. Landrum insisted she was.

When he and Holsinger left that day, he took with him yet
another surveillance tape with a section marked where Landrum had picked out the image of a woman with long blond
hair, the woman she said was Piper Rountree.

Meanwhile, in Houston that afternoon, Breck McDaniel
and Coby Kelley were back in Kingwood, standing on the
porch of the house next door to Piper Rountree's rental
house, knocking on the door. From the description Piper had
given—the house to the right as they faced her house—this
was where the woman with the daughter in Girl Scouts lived,
the woman Piper claimed could verify that she was at home

at two-thirty that Saturday afternoon, two hours before the plane from Norfolk landed. When the woman answered, McDaniel and Kelley noted that she was a Mexican immigrant and spoke limited English. As Kelley asked questions, the woman tried to answer, all the while getting increasingly upset. Kelley and McDaniel finally gave up and left.

That same afternoon, Cheryl called Kevin O'Keefe on his cell phone. She'd checked the tabs and discovered that Kevin hadn't been in the Volcano on Friday night of the previous week, only Monday and Saturday. The two started kicking around the possibilities. If that were true, then Kevin felt sure he'd seen the woman on Saturday night, since he was certain it was late in the week. But then, how could Cheryl have seen her? She hadn't worked on Saturday night.

Another element of Piper's alibi evaporated later that day when McDaniel called the neighbor woman back, this time with an interpreter on the telephone. First he assured her he wasn't investigating her or questioning her status in the United States, trying to appease any concerns she might have if she were an illegal alien. The murder case was more important than any concerns about the woman's immigration status. McDaniel explained that all he wanted from her was information on what she'd seen at her neighbor's house the Saturday before. That seemed to quiet the woman's fears.

When questioned, the woman said she had noticed a black SUV in the driveway on Saturday at Piper's house, but she didn't know what type. When asked to describe the woman who answered the door when she went to deliver the Girl Scout cookies with her daughter, the woman hesitated. The woman, she said, hadn't opened the door all the way. In fact, she stood behind it, peeking out. A baseball hat was pulled down over her forehead, and she kept putting her hand in front of her face, coughing, saying, "I'm sick."

"Was it the woman who lives in the house?" Breck asked.

"I don't think so," the woman said. "This woman was bigger."

McDaniel hung up the telephone and recounted the conversation for Kelley.

"Tina has a black SUV, too, a GMC Envoy, and she's a larger woman," Kelley said. "That could explain the e-mail to Charles Tooke and the phone call. Maybe Tina went to Piper's house to send them?"

Meanwhile, in Henrico, Kizer and the others in the Jablin task force listened to the reports from Houston and the news of the identification by Tina Landrum in the Miller Mart. They now had the information from Tarra Watford and Ray Seward's positive identification of Piper. They also knew the make and year of the van Piper had rented. Once they had that, another piece of evidence clicked into place: When he'd taken another look at the 7-Eleven convenience store surveillance tape, Jamison discovered that the woman with the long blond hair, who appeared to be Piper, could be seen in the parking lot getting out of a maroon vehicle that matched the description of the Eagle rental van.

They still had loose ends, but the investigation was clearly coming together.

Soon, a wave of back-patting erupted in the Henrico P.D.'s chief's conference room.

"All right!" one after another said, congratulating each other.

Wade Kizer judged he now had enough evidence to arrest Piper Rountree for the murder of Fred Jablin. All that remained was to decide when and how. The past few days, Piper had made herself scarce. Houston police were trying to monitor her whereabouts but hadn't seen her at her house, Tina's, or the clinic. She'd checked out of the Houstonian.

Yet Kizer wasn't too worried. He'd followed the Jablin children's custody case and knew there was a court hearing scheduled in Henrico for the coming Monday afternoon. He had a plan: He would convene a special grand jury on Monday morning. Assuming they voted to indict, police officers would then arrest Piper after she left the courtroom. Knowing what he knew about Piper, and suspecting that she was obsessed with reclaiming her children, Kizer felt certain that she would be there to demand custody.

"This is it," he said to Owen Ashman and the officers gathered around him, including Dorton, Hanna, Russell and Stem, who'd worked so hard on the case. "Great work, all of you."

Still, as Kizer walked from the meeting, he had nagging doubts. He was a careful man, not the kind who liked to walk into a courtroom unprepared, an important attribute in a good prosecutor. He couldn't help but hope there was more in the offing, more evidence that would clearly tie Piper Rountree to the crime. If she'd committed the murder—and he believed she had—it had been one of the coldest crimes he'd ever encountered, premeditatively gunning down the man who'd loved her, her husband for nearly two decades, just yards away from their sleeping children.

For all Piper Rountree's talk about how she loved her children, Wade Kizer had no doubt she'd murdered Fred Jablin with no regard for their welfare, only her own selfish motives.

15

W hile preparations were being made for Piper's arrest in Richmond, at ten the following morning, Saturday, in Houston, Breck McDaniel and Coby Kelley received a phone call.

"I'm sorry I hung up on you yesterday," Carol Freed said. "I'm ready to talk to you now."

The bulky woman with round, full cheeks, a high forehead, a straight smile, and a brash manner had been Tina's patient and friend for years. Her friends called her Cari and Tina would later describe her as a "Tina wannabe," a woman who emulated her. "She wanted to be around me," Tina said with a wry smile. "But she has a coarse sense of humor, and she's loud. We're really not very much alike."

When Freed's phone call came in, Kelley and McDaniel were in HPD headquarters working via a conference call with Houston assistant district attorney Kelly Siegler and Owen Ashman in Richmond on the search warrant for Piper's house. They offered to pick up the forty-seven-year-old Freed, to drive her to HPD headquarters to talk with them. She declined, saying she'd rather use her own car. As soon as she arrived, in early afternoon, they brought her up the elevator to the sixth floor, Homicide, and into a windowless interview room with plain white walls and bright fluorescent lights, to take her statement. At times Freed would become jittery, and they'd escort her downstairs and wait with her outside, while she smoked a cigarette, then bring her back upstairs to begin again. Throughout the hours they spent with her, Freed's mood spiraled up and down, fueled by tobacco and caffeine.

"You have to understand," she said. "This was high girlfriend drama. It was exciting."

The account Freed gave to the police began the Thursday before the murder, the day Piper boarded the Southwest Airlines flight to Houston. That day, Tina and Cari were talking when Tina mentioned she was worried about Piper. "I think she's going to do something stupid," said Tina, who described Piper as being "eerily calm."

That previous week, Tina told Cari, Piper had gone to a shooting range to target practice with Tina's boyfriend, Mac. "After he showed her how to do the beady thing [to use the gun's sites]," Freed said, Mac had told Tina that Piper was a good shot.

At the time, listening to Tina, Freed said she'd envisioned Piper kidnapping her kids and getting caught at the Richmond airport when she tried to spirit them out of Virginia. Then Tina added something else: Piper had taken off with Tina's driver's license and credit card.

"She won't get far with that," Freed said she'd responded, implying that there probably wasn't enough available cash on the card to take Piper far. Carol said she and Tina both laughed.

The next episode Freed described for McDaniel and Kelley took place on Sunday, the day after the murder, when Tina called crying and said that she needed help, something awful had happened.

"With you?" Freed asked.

"No, Piper," Tina answered.

It was then, Freed said, that Tina told her that Piper had gone to Virginia and killed her ex-husband.

Afterward, Freed said she'd gone to Tina's house, where

they sat outside and talked, Tina crying while she explained that she'd helped Piper dispose of evidence, including a purse and a wig.

"That's stupid," Freed said when she learned that Tina had thrown grocery-store-type plastic bags containing the wig and other evidence in two Dumpsters. The disposal sites didn't sound all that safe to Freed.

"We agreed to get whatever it was Tina threw away," Freed told the officers.

With that, the two friends drove to the first Dumpster, outside a convenience store. Once there, Freed said she asked Tina to go inside and buy her a Red Bull energy drink. While Tina was inside the store, Cari rummaged through the garbage bin and found a bag that felt and looked like it had a blond wig inside.

In Freed's account, from the convenience store the two women drove to a Medical Center parking lot. Tina pointed out another Dumpster, and Freed retrieved the only bag inside, which felt like it contained a purse. Apparently afraid police would be looking for her black SUV, Tina then drove Cari's Geo Metro to the defensive driving school, where she dropped her off to work. About three-thirty that afternoon, Tina returned and picked Cari up. The bags they'd pulled out of the garbage were gone, and the car sparkled. Tina had had it washed, inside and out.

The following day, Monday, the second day after the murder, Cari told the investigators she went to Tina's house again. This time her friend wasn't there, but Mac McClennahan was home. When Freed asked if he knew what had happened, Mac told her that he knew Fred was dead and that somebody had killed him. Freed then told Mac that she thought "she" was the killer, without saying who "she" was. The conversation ended quickly when Piper pulled up in her Jeep and walked toward the house.

When Piper walked inside, Cari said she did what any

friend might do, offering Piper her condolences. "I'm sorry about Fred's death," she said.

"I'm not sorry he's dead, just about how it happened," Piper responded.

Freed then told Kelley and McDaniel that she asked Piper where the gun was. Piper, she said, replied that she didn't know. "I'm your friend and we're going to get through this," Freed said she then told Piper.

Piper then brought up her computer and asked, "Do you know anyone who can change out a hard drive?"

Freed said she didn't.

There was something else that Piper said Cari could do for her: get rid of filled garbage bags in her garage. If unspoken, the implication was that there were items inside the garbage bags that Piper didn't want to fall into the hands of the police.

Perhaps still caught up in the "high girlfriend drama," Freed agreed.

With that, Piper drew a map to her house on the back of her business card and listed what items she wanted disposed of—the garbage bags and her desktop computer tower. They walked outside to the car, and Piper handed Freed her garage door opener.

With that, Cari said, she left Tina's house and drove to Kingwood. When she got to Piper's house, the bags were in the garage, just as Piper had said they would be. Freed loaded them in her trunk and then went inside the house to retrieve the desktop computer, just as Piper surprised her by walking in the door.

"I thought she'd be arrested by then," Freed told Kelley.

Out of a drawer, Piper pulled a laptop computer and gave it to Freed to take as well.

"What am I supposed to do with all this?" Freed said she asked.

"Take it away," Piper replied.

On the way home from Kingwood, Freed said she began to feel sick. The rush was evidently wearing off and the reality of the unfolding events she'd become caught up in came crashing down on her.

Later that evening, Freed said Mac came to see her at the Laff Stop, a Houston comedy club. When they started talking, he said he thought that earlier that day she was voicing the suspicion that Tina had murdered Fred. Freed corrected him, saying she thought Piper had done it.

Mac then told her that he'd talked to his sister, an attorney, and she'd advised him to consult a lawyer friend of hers in Houston, Matt Hennessy, who worked in criminal law, in the firm of Houston's well-known defense attorney Dick DeGuerin.

"You've managed to keep an arm's distance from this," Freed said she told Mac.

The following morning, Tuesday, Freed was driving down the street with her girlfriend, Patti, who she described as "a wise individual." When she filled her in on all that had transpired, Patti advised her to talk with an attorney. About that time, they saw Tina following in her car. They pulled over. Tina was crying, Freed said, and she and Tina talked. During the conversation, Tina said she'd made an appointment to talk to an attorney on Friday.

"Don't talk to anybody about any of this," Freed quoted Tina as telling her and Patti.

Freed agreed, but didn't keep the promise, instead going to her own attorney the following day. "I was not going to ride the coattails with good advice from a hysterical woman," she told the investigators. "I wanted to find out for myself where I stood."

When Freed told the attorney of her concerns, he offered to represent her but wanted a $10,000 retainer. Her best alternative, he suggested, was to take the items she'd gotten from Piper, which were still in the back of her Geo Metro, and turn them over to the police.

Instead, Freed decided to return the bags and computers to Piper.

"Where's your sister? I want to give her things back to her," Freed said she told Tina on the telephone.

As soon as she left the attorney's office, she called Patti and asked her to meet her at the Houstonian, where Tina said Piper was in Room 220. "The attorney had told me to go, do not pass go, just go," Freed told Kelley. She said she'd arrived about dusk on that Tuesday, pulling up in front of the Houstonian Hotel with Patti in her own car behind her.

"I was a mess," Freed said.

She approached a valet and asked him to help her. She unloaded the bags and the computer onto his cart, gave him a five-dollar bill, and asked him to deliver it all to Piper Rountree in Room 220. Then she and Patti left and drove to a nearby Starbucks to recuperate from the turmoil.

On Friday morning, when she went to meet with Tina to go to the attorney, Freed told Kelley and McDaniel that Betty Rountree, their mother, was also at Tina's house. As usual, Tina was running late, and when they got to the office, she went directly inside to talk to the attorney. When Tina came out, she asked Freed to talk with the man. In his office, Freed told Tina's attorney what she knew, and he recommended she retain the services of her own attorney. He picked up the telephone and called an attorney for her.

"I felt like he was putting a panic rush on me," she said.

At times, during the lengthy interview, Kelley and Mc-Daniel both had the feeling Freed was holding back, that there was more. Yet she continued to talk, adding pieces to the puzzle of the two sisters and Fred Jablin's murder.

After they left the attorney's office, Freed said she asked

Tina how much money her family intended to pitch in to help cover her legal expenses.

"We'll certainly help you," Tina said.

"I figured Tina had no idea how much it could cost," Freed told the officers. She did go to the attorney Tina's lawyer referred her to, she said, and he "scared the bejesus out of me." When she'd gotten the call that Kelley and McDaniel were looking for her, she at first wondered if it were some kind of a joke. Then the receptionist handed her McDaniel's business card, and she knew it was serious.

That morning, after she called McDaniel and before she'd come in, she said she and Mac had met at a restaurant called the Ragin' Cajun. Over lunch, Mac verified what Tina had told her earlier: that Piper was a good shot at the shooting range, once he taught her how the gun's sights worked.

Much of what Freed had told Kelley and McDaniel was damaging for Tina Rountree. Yet Freed insisted she had no malice toward her friend. In fact, Kelley realized Cari was deeply troubled by turning on Tina. When McDaniel typed up her statement, Freed went so far as to ask him to insert a comment in the report that said Carol Freed "loves [Tina] very much," and "this is far bigger than both of us."

At 3:46 that afternoon Freed signed her written statement and McDaniel notarized it. Then Freed mentioned one more thing: At one point Tina had talked about going to the police herself. Despite all she'd been willing to do to protect Piper, it appeared, Tina wasn't unaware of the possible legal consequences. Tina, after all, had a lot at stake, her nursing license and her freedom.

Listening to Freed, Kelley and McDaniel both wondered if Tina could be convinced to turn on her sister, and help their investigation.

Then, as they finished their session, Kelley put Carol Freed on the telephone with Wade Kizer, in Richmond. Kelley wanted the prosecutor to hear her story firsthand. When Freed finished talking, Kizer thanked her for coming forward and told her, "Based on what I'm hearing, we're not interested in prosecuting you. I can't speak for Houston, but if they ask for my recommendation, as long as you're cooperative there won't be any charges brought against you."

As Freed left, McDaniel was impressed with all she'd done, that she'd admitted her involvement, without shielding herself with an attorney. Now they had evidence that suggested if Tina hadn't helped plan or commit the crime, at the very least she'd been involved in covering it up.

Later, Kelley and McDaniel drove to the Houstonian Hotel, where they asked to see the head of security. As they'd hoped, the front area of the hotel, where the valet worked, had surveillance cameras recording the comings and goings. As they looked through the tapes for that Tuesday at dusk, they uncovered, at 7:11, Carol Freed driving up in her Geo Metro. They watched as, just as she said, she got out of the car, opened her trunk, and flagged down a valet, and then unloaded the bags and computer onto his cart. She handed the man a tip, and at 7:13 she drove away.

When the officers left the hotel, they took the copy of the tape with them, more evidence in what was becoming a strong case.

The day wasn't over, however. Later that afternoon another call came in for Kelley, this one from Jerry Walters. "I've been thinking, and I remember a .38 caliber gun Tina had," he said. As he talked, he told Kelley that he'd found a bullet at Piper's house, a .38 special. He'd kept the bullet and still had it in a cup in his truck.

"Do you want it?" Walters asked.

Kelley, of course, said he did.

That evening, Kelley drove to the airport to pick up Danny Jamison, the Henrico crime scene specialist working the case. The search warrant was coming together, and Kizer and Stem wanted Jamison in Houston for its execution.

In the past couple of days, Piper had called Kelley off and on, asking how the investigation was proceeding. Kelley had the impression she was trying to feel him out, to find out if she'd be arrested if she showed up at the custody hearing in Richmond on Monday morning.

"Will you be able to testify in my favor at the custody hearing?" Piper asked Kelley.

"If I'm back there, certainly I'll testify to whatever is truthful," Kelley replied.

"Okay," she said. "Good. Because I'm worried about going there."

The following day, Sunday, Kelley and Jamison refined the search warrant, working with Ashman in Richmond. McDaniel had the day off but called in regularly to help. Kelley Siegler, the Houston assistant district attorney they were working with, said she thought they could have it signed on Monday morning. The concern was that the warrant had to fulfill not only Texas but Virginia criminal law, so that whatever evidence they found and collected would be admissible at a trial.

That finished, in the afternoon the two officers drove to a sandwich place on the outskirts of Houston to meet with Charles Tooke. Piper had mentioned an e-mail to Tooke and a call from her the afternoon of the murder as part of her alibi. She was eager for Tooke to talk to the police, and just that morning she'd called Charles saying, "Coby Kelley, the Virginia investigator, wants to talk to you about when you talked with me that week."

"Okay," Charles agreed.

For his meeting with Kelley and Jamison, Tooke brought

a copy of the e-mail from Piper that had arrived at 3:29 that Saturday afternoon. It read, "I've got a brief question for you." Tooke had also printed a copy of his response for Kelley: "I'm a boxer shorts. Not briefs."

He also had with him a copy of his cell phone bill, showing the time he'd received a voice mail from her that same afternoon, the day of the murder. Although she said again that she had a question for him, when he called her back, he told them, she failed to respond.

Hard of hearing, Tooke asked them to repeat themselves often as the conversation progressed. He believed Piper was being wrongly suspected and that she'd had nothing to do with the murder, and he was eager to assist the police.

"I kind of had a thing for Piper," Tooke admitted. "But don't tell her."

As they listened, Jamison and Kelley heard nothing from Tooke to convince them they were looking at the wrong suspect. What was clear was that Piper wasn't at work, answering Tooke's calls, on either Thursday or Friday of that week. The last time he'd talked with her on the telephone, although they routinely conversed almost daily, had been at 12:45 P.M. on the Wednesday before the murder. As to the e-mail and voice mail on the day of the murder, while police believed she was in the air flying back to Houston, those could have been sent by anyone using her e-mail account and a tape recording of her voice.

Something Tooke said did, however, interest the two investigators: On the call from Piper urging him to talk with them, Tooke said that she'd mentioned she was on her way to the airport, to fly to Virginia to attend the following day's custody hearing.

After they left Tooke, Kelley called Wade Kizer. "She's flying in today, and she'll be at the hearing tomorrow," he told him.

"We need to find her," Kizer said.

For the rest of that Sunday, officers in Richmond attempted to find out where Piper Rountree was staying in Richmond. They wanted her under surveillance so that as soon as Kizer had a signed indictment from the grand jury, they could quickly move in and make an arrest. They called hotels and motels, but came up dry. Officers staked out the airport and didn't find her. As Monday morning broke, neither Kizer nor the police knew where Piper Rountree could be found.

That Monday morning David Ferguson, McDaniel's partner, was on a long anticipated vacation, hunting in Mississippi. It was one of his favorite things. He loved getting the deer in his sights, lining up the rifle, and then slowly squeezing the trigger. Yet, this day, he was having a difficult time keeping his mind on the woods and the deer, instead wondering what was going on in Houston. When he called in about 9:00 A.M., he talked with his partner.

"We've got the search warrant," McDaniel said. "We're going in this afternoon, after they arrest her in Virginia."

"Hope you have better luck with your hunt than I've had with mine," said Ferguson, who hadn't seen a buck all day.

On the warrant were the items the police would be searching for when they entered Piper's house: ammunition, wigs, Tina's identification, bills, bank and cell phone documents, computers, luggage, clothing and cell phones. Among the justifications for the warrant were listed all the reasons Piper Rountree was a suspect, including the purchase at the CVS pharmacy of size small latex gloves.

"Your affiant has found that criminals wishing to conceal their fingerprints wear gloves," McDaniel had written.

It was a busy morning for McDaniel, who had a second document signed by a judge: a warrant for Tina Rountree's arrest. Virginia didn't have laws to punish those who tried to cover up a crime for a family member, but Texas did, and the pocket warrant—so named because it was theoretically kept secret until removed from a pocket and presented at the time the arrest was made—charged Tina with a felony, tampering with evidence.

"All the tiny puzzle pieces were fitting together," Owen Ashman says of that Monday morning. Sometimes the pieces of evidence were small, seemingly insignificant, but when joined with all the other evidence, they formed a persuasive picture. One such fact was something Ashman noticed on the Homestead Suites registration form. Instead of her rented van, a Ford Windstar, Piper had listed her vehicle as a Chrysler Voyager, the type of van she'd received as part of her settlement in the divorce.

All around, those involved were starting to feel better about the Rountree case, including Captain Stem. Despite the spats he and Kizer had over the case, that morning he had to admit he felt better about the evidence they'd collected than he would have if they'd made the arrest the previous week.

To set events in motion, before the custody hearing started that morning, Owen Ashman went to see the family court judge, to explain that Wade Kizer was, at that very moment, in front of a grand jury, and that he hoped momentarily to have an indictment ready to arrest Piper Rountree for Fred Jablin's murder if not before the custody hearing began, immediately after she left the courthouse.

In the grand jury's meeting room, Kizer and Kelley's partner, Robin Dorton, outlined the evidence that they said proved their case, all the documents and witness statements collected by Hanna, Kelley, and the others working the case. When they finished, the jurors voted to issue two indictments that charged Piper Rountree with unlawful use of a firearm and first degree murder, the victim one Fredric Jablin.

At that point, late on the morning of the custody hearing,

with the signed indictments in Kizer's possession, all that remained was to find Piper Rountree and make the arrest.

While the others were on the lookout for Piper in Virginia, in Houston, Kelley and Jamison, aided by local officers, were inside Piper's house in Kingwood. Some of what they'd hoped to find was gone, including the two computers. But then, after talking with Carol Freed, Kelley would have been surprised to find them. Where the computer tower had been, the cords dangled untethered. Jamison took a photo of the gutted computer desk as evidence.

The officers didn't walk out empty-handed, however.

As a result of the search, Jamison collected envelopes full of Piper's documents, including copies of statements from the Jerry Walters Wells Fargo account, Piper's black day planner, and two cell phones. He confiscated hotel-style miniature toiletries, hoping to tie them back to the Homestead Suites, two digital cameras, and a red carry-on bag and a black suitcase, fitting the description of the luggage witnesses said the woman they'd identified as Piper Rountree carried in Virginia the week of the murder.

However, it was the discovery a Houston detective made in the kitchen that most excited Jamison and Kelley. There, tucked inside a black beret wrapped inside a ball of dirty clothes, the detective found the red wig purchased on the wigsalon.com website. Except for the color, paprika, it was a twin to the Frosti Blond Stevie wig that, like the computer, had vanished.

At the University of Richmond that afternoon a memorial service was held for Fred Jablin in the Robins Pavilion of the Jepson Alumni Center. Two hundred fifty people attended, many of them Fred's students, former students, colleagues, and peers from across the country.

A cantor from the synagogue sang, and at the podium Joanne Ciulla talked of the man whose office had been next to hers for a decade. Many chuckled knowingly when she described the stacks of papers and journals in Fred's office. John Daly, who'd flown in from Austin, recounted his memories of his old friend, the Fred who at Purdue studied under a green glass shade, intent on making his mark on the field. And he'd done so. In a letter read to the audience, Linda Putnam, Fred's coauthor on the two books that set the standard in organizational communication, called Fred a dedicated scholar, a mentor of students, and a pioneer in his field. "Many of us in the field . . . feel as if we have lost part of our own lives," she wrote. "He will be deeply missed for his dry wit, his serious dedicated pursuit of knowledge, and his friendly manner."

As before, Michael Jablin and his wife were there with their two children and Jocelyn, Paxton, and Callie. During the ceremony, one of the professors walked down from the stage and handed each child a stuffed animal with a sash that read, SOMEONE AT UR LOVES YOU.

Meanwhile, in the parking lot at the Henrico County Courthouse, down the street from the Henrico P.D. headquarters, Chuck Hanna was watching for Piper to drive up when he heard via radio that she'd already entered the courthouse vestibule accompanied by her brother, Bill, and another attorney. Minutes later, inside the designated courtroom, Owen Ashman sat in the gallery as Piper entered, wearing a somber black dress with a large silver crucifix on a chain around her neck.

How manipulative, Ashman thought.

From her vantage point, Ashman could see Piper and her brother with her attorney. When Ashman assessed Piper, she thought her face looked cold and chiseled. Yet when she rose to talk to the judge, Piper's voice sounded frightened and vulnerable. Since Judge Hammond, the same judge who'd ruled in the divorce case, knew Piper would probably be arrested before the day was out, the hearing was nothing more than a formality. Twenty minutes later, after Piper's attorney argued that with Fred dead the children should be turned over to their mother, the judge ruled: The children were safe with their aunt and uncle, and at least for the time being would remain with Michael Jablin, as Fred had asked in his will. Once the criminal investigation concluded, however, another hearing would be set to determine permanent custody.

As she listened, Piper's shoulders sagged.

Moments later Ashman followed Piper and her entourage from the courtroom and out the courthouse front doors, where they brushed past the media and rushed to their car.

Minutes later they pulled out of the parking lot.

The waiting police, including Hanna, allowed Bill Rountree to drive off with Piper beside him before following in four unmarked cars. As they drove, the police cars moved into position: one car in front, one behind, and one on each side. When Bill stopped at a light on the intersection of Stillman and Broad Streets, Hanna and the other officers had him boxed in. They turned on their lights and sirens. It was what's called in police parlance a "dynamic vehicle takedown." When Hanna looked over at Piper, she appeared starkly white, the blood drained until her skin shone an unnatural white. Before even being told to, while still in the car, she put both hands up in the air.

Hanna and the other officers exited their cars, weapons drawn, and opened the car doors. Piper and Bill climbed out.

"What's going on?" Bill asked.

Piper said nothing. Hanna looked at her, all in black with

the large silver crucifix dangling around her neck, and said, "Ms. Rountree, you're under arrest."

She didn't ask any questions, including what she was being charged with. Instead, Piper lowered her head and cried softly. Hanna couldn't help but think she looked defeated.

"Look, there's a lot of news media around," he said. "You can get in my car and we'll drive you back to headquarters to talk there."

"Okay," Piper whispered. "Thank you."

Minutes later, in a second-floor interview room, Piper Rountree was read her rights.

"I want my attorney before I talk with you," she said.

"Who's your attorney?" Hanna asked.

"Murray Janus," Piper said, naming the man many considered the dean of criminal law in Richmond.

An hour and a half later, Janus, a slightly built, genteel man, arrived and went in to talk to Piper. When he emerged, he walked back up to Hanna. "Ms. Rountree will not be making any statements," he told him.

"That was that," says Hanna. "From that point on, she never talked with us again."

In Houston, the search warrant had been fully executed and the Kingwood house was secured. Meanwhile, Coby Kelley and Breck McDaniel were on their way to the Village Women's Clinic to arrest Tina Rountree on a felony charge of tampering with evidence. Inside the clinic, they asked the receptionist to get Tina. Moments later she came out, looking not at all surprised.

"She acted like she knew we were coming," says Kelley.

They allowed her to take off her jewelry, including her Rolex watch, and leave it with the receptionist, and then walked her outside and put her in the back of a waiting squad car. At Houston P.D. headquarters, they brought her to the same interview room where two days earlier they'd questioned Carol Freed.

"Listen, you're here for a reason. We have evidence," McDaniel told Tina. "You're not the main player, but what you did was wrong. If you're found guilty, you have a lot to lose, including your state nursing license. You could avoid all this. If you have evidence that can help us, you want to cooperate and get ahead of this."

Tina looked at them, a nonchalant look on her face, appearing not the least concerned. "I want to speak to my attorney," she said.

When the attorney arrived, Breck explained again that it wasn't really Tina police wanted, and that they'd be interested in working out a deal with her if she was cooperative and would supply evidence and testimony against her sister.

"This is a third degree felony," Breck told him.

The attorney went in and talked to Tina, then came out and said his client had nothing to say.

Investigator McDaniel handed the attorney his card.

"We have a lot of evidence. Every day that goes by we'll need Tina less and less," he told him. Warning him that Tina needed to cooperate while she still had something to offer, McDaniel told him what he'd told Tina: "She wants to get ahead of this thing."

"My client may talk to you, but not today," the attorney answered.

With Tina in jail, Kelley and McDaniel decided it was a good time to approach her boyfriend, Mac McClennahan. They called him, and when he answered the phone, they said they wanted to have a meeting. Mac replied that he wanted to call his sister, the lawyer. She was the one who called back. "Mac would like to talk to you," she said. "But my friend from law school is going to be there with him."

* * *

That night Mel Foster watched a television news segment about Piper's arrest, accompanied by clips of her in her somber garb with the heavy silver crucifix around her neck, leaving the custody hearing. Mel couldn't help but think back to when the Jablins had first moved in behind her. Piper had always been eccentric, but they'd seemed happy. Who could have predicted any of this would have happened? she wondered.

In Houston that same night, at Charles Tooke's house, the telephone rang.

"Hello there," Charles said. "What's new with you?"

"Well, I've been arrested," a glum Piper said. At that, they were both silent.

16

T hree days after Piper Rountree's arrest, on November 11, Kizer, Ashman, Janus, his associate Taylor Stone, and Piper were all in front of a judge, discussing the possibility of bond. The formal charges were read: first degree murder and the use of a firearm in the commission of a crime.

In the courtroom, Janus made a passionate argument, saying that Piper was the mother of three children who needed her even more after the loss of their father. The Christmas holidays were coming up, and she should be with them. He asked the judge to give her a high bond and force her to surrender her passport, to order her to stay in Virginia and not return to Texas until the trial. Janus talked of her medical problems, including the removal of her spleen as a child, which left her vulnerable to infection while in the county jail.

"Ms. Rountree's children are three good reasons why she's not going anywhere," Janus pleaded.

"This woman has no ties to Richmond," Kizer said when it was his turn. She'd hid from law officers, taking a room at the Houstonian, when they wanted to talk with her. He recounted the crime, stressing how Piper had disguised herself as her sister and used false identification. What was there to stop her from disguising herself again, this time to flee Virginia, even the country, and not return for trial?

Before the hearing ended, the judge had ruled in favor of the prosecutors. Piper Rountree, he said, was not eligible for bond and would remain in jail until the trial, scheduled to begin February 22 of the coming year, 2005.

In Richmond and throughout the state, Murray Janus was well known. Voted by the *Virginia State Business* magazine as the state's top criminal defense attorney, he was considered a true Virginia gentleman, a senior statesman of the Richmond courts, and a refined man who collected art and supported local charities. Although slight in stature, in the courtroom Janus was a strong presence. Juries liked his straightforward approach and his sometimes self-deprecating sense of humor, and other attorneys admired him. Even the Henrico commonwealth attorney respected Janus. "He's a tough but a fair opponent," says Kizer. "We have no quarrel with him."

Over the years, Janus had often been called on to represent Richmond's elite, and his cases were legendary, from ABSCAM to the Dalkon Shield litigation. In many ways, Janus and Fred Jablin were alike. They both had brilliant minds and had risen to the top in their fields. Like Fred, Janus came from modest beginnings. In Janus's case, his parents ran a Richmond variety and hardware store. Neither Fred's nor Janus's parents had gone to college. In fact, Janus was the first in his family to go on to higher education. Also like Fred Jablin, Murray Janus began his career studying under a powerful mentor: Robert Merhige Jr., his senior law partner, who went on to become a judge and preside in the 1960s over the desegregation of Richmond's schools. And like Fred, Murray Janus loved his work. "I've never had a boring day yet," he says. "Every time I think I've seen it all, done it all, something new comes through the door."

What came through his door in November 2004 was an offer from Piper Rountree's family, who were willing to pick up Janus's substantial retainer to have him defend her. Piper had tried to hire Janus as her attorney before, in the midst of her divorce, and he declined. Janus didn't like to take divorce cases other attorneys began. This time, with Rountree facing a murder charge, he accepted.

To help him on the case, Janus enlisted the aid of Taylor Stone, a young associate in his firm. Like Janus, Stone, a dark-haired, intense man who'd left a six-year career as a stockbroker to enter law school, was a Richmond native. A distrust of authority had persuaded him to practice law. In college, Stone had been a Deadhead, a follower of the rock band the Grateful Dead, and he'd retained a bit of an edge from those years, never quite trusting police. "I get turned off by twenty-two-year-old kids wearing badges," he says.

When he graduated from the University of Richmond law school, Stone applied to work with Janus for one reason: "He's the best in the business." He found Janus fatherly, yet tough. "When he gets a client off and that client messes up again, he gets upset with them," says Stone. "He wants them to turn around their lives."

When it came to Piper Rountree, Stone had heard about the Jablin case on television and wasn't surprised when Rountree ended up on their client roster. "Working with Murray, you get used to seeing cases in the headlines and then having the accused walk in the firm's office door," he says.

The Piper Rountree case would be a trial by fire for Taylor Stone, the first murder case he'd help defend.

Two days after Piper's arrest, Stone had his initial meeting with the client. He and Janus formed a similar opinion. "When you meet Piper Rountree, you think, this woman's not capable of murder," says Janus.

In Houston on Tuesday, Tina Rountree remained in jail, waiting to make bond. Meanwhile, Kelley heard that Mac had been on the news, talking about the Rountree case. Unhappy,

Kelley called Mac. "You wouldn't talk to us, but you were on the news?" he said.

"I wasn't on the damn news," Mac said.

Later, Kelley found out it was actually Marty McVey who'd given the interview, but Kelley didn't back down, cautioning Mac that he could have exposure on the case, that unless he cooperated they might consider charging him as an accessory.

That afternoon, Kelley and McDaniel sat down with Mac and his attorney, Matt Hennessy, a tall slender man with a mop of thick dark hair and a hawklike gaze, at his law office. Throughout the interview, Mac appeared nervous, as if not eager to help but not willing to lie to cover anything up. The two officers prodded him, asking questions about his relationship with Tina and what he knew about Piper and the events of the preceding weeks.

"Piper said it would be fun to go to a gun range," Mac told them. Mac had a concealed weapons permit, and he'd mentioned to Piper while they were working in Galveston on the Tuesday before the murder that he'd wanted to go to the firing range. She said she'd like to go with him. After work, she'd dropped him off at Tina's house, where he picked up his handgun. Then he drove and met her, as they'd planned, at the 59 Gun Range, off I-59 north, between Houston and Kingwood. Inside, they'd signed in, rented a .22 and purchased a box of .22 shells for Piper, along with a box of ammo for Mac's .40 caliber semiautomatic.

On the gun range, Mac helped Piper practice lining up her shots by using the gun's sights, and they stood side by side, aiming and shooting at targets. Then, after Piper had shot half of the box of .22 shells, she suddenly left. When she returned, she had a different gun: a snub-nose .38. When Kelley asked where the gun had come from, Mac said he didn't know, but he'd assumed she'd gone inside, returned the .22, and rented the .38. She also had a box of fifty .38 shells.

From that point on Piper fired the .38, working her way through two-thirds of the ammo. When she'd finished, Mac tried out the .38 and used up the rest.

The following day, Wednesday—the day before Piper flew to Richmond—Mac explained that he was at Tina's house when Piper walked in complaining about being sore from shooting the guns the day before. Then, without explaining why she wanted it, she asked Mac for a bill with Tina's address on it. He told her to take any she wanted; he had a pile of them on the table.

"She took the cable television bill," he said.

That bit of information fit into what Kelley and McDaniel already knew from the investigators in Richmond: Piper had presented Tina's Time Warner cable television bill as proof of address when she rented the Ford van at Eagle.

Although he'd been working with her earlier in the week of the murder, training as a landman, Mac didn't see Piper either Thursday or Friday of that week. She'd told him she was attending a law conference in Fort Worth.

That weekend, Mac visited his parents in the Hill Country. When he returned to Tina's on Sunday, she told him about the killing. It was when he went to McVey's office to pick up Tina, while Kelley and the other officers waited for her at the house, that Mac saw Piper for the first time after the murder. He took that moment to offer his condolences over Fred's death.

"I love you," Piper whispered in his ear. Then she said, "Please don't tell the police about the firing range. It would complicate things."

Throughout the interview, Kelley asked questions and Mac supplied answers.

Sometime during that week before the murder, Tina told Mac that she needed her passport to use it for identification. Her driver's license and credit card were missing, and she said she thought Piper had taken off with them.

"Why?" Mac had asked her.

"I don't know," Tina replied.

As Kelley and McDaniel listened, Mac recounted the time, a year earlier, when Tina had a .38 caliber pistol she kept in her house. When he got ready to break off the relationship, he gave the gun to Piper and asked her to get it out of the house. He didn't want the gun there, he said, because he didn't know how angry Tina might become.

"Was it the same gun as the one at the gun range?" Kelley asked.

"No," Mac said.

There was something else Mac thought they should know: Early in the week after the murder, Tina had called him from the Houstonian, wanting him to bring her computer to her. He had refused.

"I was uncomfortable with that," he said. "I told her she could do anything she wanted, but I was not getting involved."

Later the computer was gone from the house, and he believed Carol Freed had taken it to Tina.

"I told Cari to be careful. She told me, 'I'm all lawyered up,' " Mac said. Then, assessing his situation, he added: "I was there while things were going on, but I made a point of not asking."

Carol Freed hadn't told Kelley and McDaniel that she'd removed Tina's computer.

After the officers finished interviewing Mac, Kelley called Freed and inquired about the computer incident. Sounding nervous, she explained that Tina had called her earlier that week and asked her to take her personal computer out of her house. "There's nothing on it you need to worry about," Tina insisted, saying the hard drive held only personal matters, including her financial files, but that she feared the police might come with a search warrant and seize it.

As before, Freed did as her friend requested. That day at Tina's, the housekeeper helped Cari unplug the computer and load it into her car. Mac was there, she said, but refused to help, and told her, "Just don't tell me what you're doing."

That night, Freed kept Tina's computer in her car, but it gnawed at her, making her feel as if she were "getting in deeper and deeper." She also mentioned that when Tina came to pick it up the next day, she was angry at Piper, who she described as drinking heavily. To Kelley, Freed remarked that she thought it odd that Tina was furious with Piper and yet risking everything to protect her.

Later that afternoon, Kelley and McDaniel checked on another of Mac's accounts from the week of the murder, at the 59 Gun Range, a grim stone building off Interstate 59. On the lot next door, adjacent to the parking lot, dilapidated double-wide mobile homes sat as if abandoned, windows broken and the vinyl siding scarred by graffiti. On the gun range door was a warning sign: ENTER AT YOUR OWN RISK.

Inside, it was cramped and cluttered. McDaniel and Kelley stood at the dirty glass-topped counter, where customers bought ammunition and rented blue plastic ear protectors and goggles, then walked through double doors into the range itself, where they shot in stalls the length of bowling alleys. The floors were covered with spent brass cartridges, and the paper targets hung from a rope line with clothespins were headless silhouettes of a human torso with a bull's-eye drawn on the chest.

"Nice place to go to get ready to commit a murder," Kelley whispered to McDaniel.

One clerk came over to help them, and soon a few had gathered, listening as they showed their badges and explained why they were there. When Kelley showed Piper's photo, none of them said they remembered her, but one clerk looked through the records and found the form Mac had filled out when he checked in at 7:07 that evening. Only his Texas driver's license was copied onto the form, but it was signed by both Mac and a woman, a woman who'd initialed the blanks *PR* and signed what appeared to be "Piper Rountree."

One thing struck Kelley on the form: While it showed the rented .22 and the three boxes of ammo they'd purchased, he couldn't see where anyone had noted that Piper had rented a .38.

"That's because we didn't rent her a .38," the clerk said.

"It was her gun?" Kelley asked.

"I guess so," the clerk said. "All I can tell you is that she didn't rent it from us."

It would turn out to be a good day for the investigators. First, they'd uncovered evidence that Piper had a .38 caliber weapon, like the one used in the murder, not only when she checked in at the airport but days earlier, when she'd practiced at the 59 Gun Range. Then, Kelley talked with Kevin O'Keefe, and Piper's Volcano alibi sizzled out.

Earlier that day, Cheryl and O'Keefe had again compared notes. O'Keefe was by then certain he'd seen Piper on Saturday night, but that didn't jibe with Cheryl's schedule. She hadn't worked that Saturday night. But then something occurred to Cheryl that settled the matter. Finally, it had all begun to make sense.

"Oh, I know what happened," she said, explaining that although she hadn't worked that Saturday night she had been in the Volcano that evening. She'd gone there to meet friends to go to a Halloween party.

"We must have both seen her Saturday night," O'Keefe said, and she agreed.

When O'Keefe talked with Kelley, he explained what they'd discovered. "It wasn't Friday, it was Saturday night," he said. "It had to be Saturday night."

By then O'Keefe had remembered something else that made him even more certain: the patrons in the bar the night Piper came in were wearing Halloween costumes. In fact, he'd worked that Saturday and come in directly afterward, still wearing his overalls. Some of the other patrons had teased him, saying it was an odd Halloween costume. That meant it had to be the night before Halloween.

Kelley thanked O'Keefe then hung up.

Afterward, Kelley thought about what he'd just heard and how cold-blooded it was: Fourteen hours after her ex-husband's murder, Piper Rountree had calmly walked into the Volcano to set up an alibi.

With Piper in custody and the evidence against her piling up, Kelley's work in Houston was coming to an end. He and Jamison had shipped the evidence they'd collected—including records from the 59 Gun Range, the items they'd taken from Piper's house with the search warrant, and the surveillance tapes from the Houstonian—to Richmond. Jamison wasn't scheduled to leave until Thursday, but that Wednesday, eleven days after he'd flown in, Coby Kelley flew out of Houston Hobby to Richmond. However, that didn't mean Breck McDaniel was off the case.

That day, while Kelley was in transit, McDaniel went to an Academy sporting goods store near Kingwood, on the hunch that he might be able to find out if Piper had purchased a gun. He came away empty-handed. Although the store filed the required federal forms, he was shocked to discover that they weren't able to search the database by name, only the gun's serial number. Without that, they had no way to access the federal records.

Undaunted, McDaniel decided to play a second hunch and returned to Hobby Airport. Once there, he tracked down the office in charge of the airport parking garages. His hope was that the facility had surveillance cameras near the entrances and exits. If it existed, he wanted a video of Piper entering or leaving the garage. Instead, he discovered that each day an attendant walked the garage with a computerized handheld wand, recording license plate numbers. The airport kept records, documenting what days the cars were in the garage, to determine what to charge customers who'd lost their parking tickets.

"Will you look for this car for me?" he asked, handing them a description of Piper's black Jeep Liberty with the license plate X54-JBJ. They agreed, and he left.

Later that same day, the parking manager at Hobby called. He'd found a record of that license plate number. Piper's Jeep, he said, had been parked at Hobby from Thursday through Saturday, the same days the Southwest Airlines tickets showed she was in Richmond.

In Houston on Tuesday, Tina was released from the Harris County Jail. Her friend, Glenda King, would later say that Tina emerged sobbing, saying over and over, as if in shock, "I didn't do anything, I didn't do anything."

Meanwhile, in Richmond, Owen Ashman had begun compiling the evidence to turn copies over to Murray Janus, all the bits and pieces they were pulling together. Discovery laws required that the defense be informed of the evidence against the accused, and Kizer and Ashman had no qualms about showing Janus just how difficult a case this was going to be. When the first folder arrived in Janus's office, it was thick with bank and cell phone records, along with witness statements. Still, the prosecutors didn't have everything spelled out, and, even before Janus was able to decipher all the cell phone records, he learned from the *Richmond Times-Dispatch* that Piper's phone had been used in Richmond the weekend of the murder, and that witnesses would testify it was her voice on the telephone.

"This wasn't going to be good for our side," Janus says with a grimace.

Tina, too, it would later appear, was beginning to understand how dire the situation was. Telephone calls in and out of the Henrico County Jail were recorded, and when Ashman listened to the two sisters talking the week after the arrests, she had the impression Tina was beginning to piece together the evidence against her. Without naming her, Tina indicated she believed Carol Freed had turned against her.

"That's great," Piper said, apparently missing the point.

"No, that's not great," Tina countered.

In the Henrico courthouse, Coby Kelley combed through the three large file boxes of Jablin-Rountree divorce records. What he found was that Fred had been meticulous, showing up at every hearing, filing every paper, supplying evidence that included Piper's own e-mails for the court to show that she'd been unreliable with the children. Piper, on the other hand, although an attorney, had missed deadlines and not shown up for hearings. Throughout the records, it was easy to see how acrimonious the divorce had been. If there were any doubt, the forty-two-page psychological assessment of Fred by Tina clearly showed the hate the two sisters had for Piper's ex.

Meanwhile, in the Henrico jail, with time on her hands, Piper wrote letters to her children at Michael Jablin's house in northern Virginia. She'd been trying to call them, she said, but she assumed no one had told them of her messages. She included instructions on how to accept a collect call from the jail, and said that she understood their world had been dealt a horrible blow. Yet, she contended that she didn't understand why she was being charged, and even claimed that the police had told her she'd been cleared.

In closing, Piper said she was close to them with every breath she took, and that, in her sleep, she sent her spirit to be with them.

There was no doubt that it must have been a confusing time for all three of the children.

At the end of November, Owen Ashman went to the Boyds' home, and Michael Jablin brought the three children there to meet with her. Immediately, Ashman was struck by how beautiful the children were and how quiet, although when she arrived, they were with their friends in the kitchen, making cookies.

"I'm sorry about your father," Owen began. "But there are some things we need to talk about. They may not seem important to you, but they could help us find out who did this."

What the prosecutor didn't say was that the evidence the children had to offer could help prove their own mother had murdered their father. That wasn't something Ashman wanted the children to consider. And it wasn't really true. Despite the potential damage their testimony could deal their mother's defense, the children weren't actually testifying against her, only telling the truth as they knew it.

That afternoon, Paxton was matter-of-fact, saying his mother had called him that Friday while he was playing poker with his friends in one boy's garage. The poker craze was on across the country, especially on television, and the twelve- and thirteen-year-olds had been betting for nickels and dimes. Piper had talked not only with him but with his friend, Russell Bootwright.

Throughout the conversation, none of the children asked Owen questions, although their minds must have been reeling with them: Did our mother kill our father? Or, who killed our father? Owen understood. She knew they wanted their lives to go on and be normal, just as she and her brothers had wanted after her own father's death.

"We didn't want to talk about it, either," she says.

By early December, Wade Kizer had brought on his right-hand man to work on the Rountree case, the chief deputy commonwealth's attorney, Duncan Reid, an angular man with dark hair graying at the temples, a salt-and-pepper beard and mustache, and a passionate energy. In his twenty-eight years as a prosecutor, Reid had never gotten over the thrill of walking into a courtroom. Unlike many attorneys, he enjoyed trying a case in front of a judge and a jury. "I like being on the right side," he says. "If I don't have the evidence, I drop the case before opening arguments begin."

When he looked at the Jablin case, Reid thought he understood how Fred had become so immersed in fatherhood. Happily married for thirty years, Reid was like that with his own children, taking them to intellectual competitions, swimming and soccer, working around the house with them and riding bikes. "Being a parent is a hard job, and I think the world of people who are single parents and do it well," he says.

The week before he began on the Rountree case, Reid had tried a home invasion case where a gun had been put to the head of a two-year-old. Two people had been killed. After it was over, Kizer had promised Reid a break, but when he considered Reid's talents, he recanted the promise and asked him to work with Ashman on the Jablin case. The talent he wanted was Reid's ability to interpret cell phone records.

With all the cases his office had pending, Kizer considered the Rountree cell phone records a daunting task. They were two and a half inches thick, single spaced, with thousands of calls. Kizer's payoff for enlisting Reid's services came quickly. One of the first days he examined the records, Reid singled out the 411 information calls on the bills. One, he noticed in particular, made on the Wednesday before the murder, had an unidentified number called immediately after it. When he dialed that number, the person who answered said, "Sportsman's Outlet."

"What kind of a place is this?" Reid asked.

"A gun range," the man said, and in Richmond, Duncan Reid smiled.

* * *

Breck McDaniel walked into the Sportsman's Outlet in Houston after Coby Kelley called from Richmond and asked him to investigate the lead. Minutes after McDaniel arrived, the clerk had gone through the records and found that shortly after Piper called on her 7878 cell phone, at 7:11 P.M. on the Wednesday before the murder, a woman checked in at Sportsman's Outlet and showed identification saying she was Tina Rountree. "Tina" had purchased a box of .38 special bullets for $10.99 and paid a ten dollar practice fee. What she didn't do was rent a gun.

"She brought the gun with her?" Breck asked.

"Well, she didn't rent one here," the clerk told him.

No one at the gun range could identify the woman in the photo, but when Kelley went in to tell Captain Stem what McDaniel had discovered, Stem was pleased. "Despite being an attorney, Piper Rountree left a lot of bread crumbs lying around. She thought she was smarter than the rest of us," he told him. "Too bad for her when she finds out she's not."

Kelley had described the Volcano bar and the cast of characters in Houston for the prosecutors: Piper's lover, Jerry Walters, the cowboy oilman with the gruff voice; Carol Freed, the brash comic; Mac, the rock and roll DJ and Tina's longtime boyfriend; and Charles Tooke, the guy who'd had a crush on Piper. In early December, Owen and Kizer flew with Kelley to Houston, where they'd be able to personally size up their future witnesses.

Their first day there, they met with Carol Freed at HPD headquarters, and Wade presented her with an immunity agreement to sign, one that said if she cooperated and testified truthfully about all she knew, she wouldn't be prosecuted as an "accessory after the fact to the murder of Fred Jablin."

The woman Tina called "Cari" had brought two things

with her to give them as well: the business card Piper had given her with a hand-drawn map to the Kingwood house on the back, and a purple long-sleeve T-shirt Freed said belonged to Piper.

"I forgot to tell you," she said to them. "Piper gave this shirt to me and asked me to get rid of it."

Looking at the shirt, they wondered what forensic evidence it might offer. They immediately bagged it and had it shipped to Henrico for analysis.

Kizer then told Freed that he wanted her to show them the places where she and Tina had retrieved the garbage bags. Following her Geo Metro with Breck McDaniel, in his unmarked car, the prosecutors and Kelley drove through Houston until they hit a point where McDaniel questioned where Freed was taking them.

"We're not going the right way," he said.

Kelley dialed Freed's cell number. "What's up?" he asked.

In the backseat, Ashman and Kizer were chuckling about something unrelated to the case, but Kelley surmised from the tone of her voice that Freed thought they were laughing at her.

"I know," Freed said, sounding angry. "I'm making a U-turn."

Minutes later they pulled into a Texas Medical Center parking lot and drove up the entrance ramp, until Freed pulled up in front of a Dumpster. With that, they all got out of their cars.

"You really need to work on how you talk to people," Freed, still furious, said to Kelley.

"Cari, have I done something to offend you?" he asked.

Seething, with no explanation, Freed got back in her car and drove off.

"Did I say something to offend her?" Kelley asked, turning to Kizer.

The prosecutor shrugged.

Freed returned minutes later and got out of the car, calmer. "I know I'm in a bind," she said. "I need to cooperate."

Ashman knew she was still furious. The prosecutor saw the entire episode as odd, as if Freed had just "snapped."

"Cari, is everything okay?" Ashman asked. But she didn't answer.

An hour or so later they were in the convenience store parking lot, looking at the second Dumpster, when Freed snapped again with no explanation, driving off and leaving them behind.

"Let's give her a little time," Owen said.

But this time Carol Freed didn't return.

The first time Kelley went to the 59 Gun Range, no one volunteered that they'd remembered seeing Piper. This time, with Wade and Owen beside him, the response from one of the men behind the counter was markedly different.

"I remember that woman, the one you were showing the picture of last time you were in here," said one of the firearms instructors, Boyd Adams. "Someone else was helping her and the guy she was with at first, and then I went over and helped them."

Kelley was skeptical. Why would Adams suddenly remember now when he hadn't earlier? Then the man said something that got all their attention. "I noticed her because she was really pretty, and we don't usually get attractive women in here," Adams said. "And because that woman said she had her sister's identification."

When Kelley again showed Boyd the photo of Piper in the red dress, he immediately said, "That's her. I'm ninety-five percent sure."

"Did she rent the .38 she shot?" Kelley asked.

"No," Adams said. "She walked outside and came back in with that. She bought some ammo for it, but it was her gun."

* * *

When Wade Kizer called Jerry Walters to set up a meeting, Walters was on his way to Victoria, Texas, 130 miles south of Houston, for work. He'd just gotten a call that his ten-year-old bloodhound, Bertha, who he'd spent thousands on to treat for bone cancer, had died. "Come if you have to. If you can find your way to the La Quinta," he growled.

Before they left Houston, however, the phone rang. Carol Freed was calling. "I'm really sorry," she said. "I'll be properly medicated in the courtroom, when I testify."

They met her at a small Mexican chain restaurant called Chipolte. As soon as Freed saw Kelley, she walked up and gave him a hug. "I'm so sorry," she said.

"I'm sorry I hurt your feelings," he responded.

"No, no," she said. They talked for about an hour, Freed confirming for Wade and Ashman all she'd told Kelley.

Later that day, at the La Quinta Hotel in Victoria, Texas, Jerry Walters wasn't in a good mood. He hadn't looked forward to talking to the investigators anyway, and now Bertha was dead and he'd been up all night driving from Baton Rouge to get there in time to work that Monday morning. He was tired and out of sorts.

They talked, Jerry explaining what he'd told Kelley about the bank account and the cell phone calls. The prosecutors both wondered if he'd held back information. There was that phone call, eleven minutes, to him from Piper on the day of the murder. Had she confessed to him and he wasn't admitting it?

"I don't remember that phone call," Walters told them, just as he'd told Kelley. "I'm sorry, but I can't even tell you if we talked or not."

The next day, at Hobby Airport, Allan Benestante, the security agent, and Kathy Molley, the ticket agent, both easily identified the woman they'd talked with as the woman in the photo again. By then Benestante had seen Piper on television, and he felt more certain she was the woman who checked the gun. And at the Volcano, Kevin O'Keefe told the prosecutors the strange story of Piper's visit there that ended with her bringing in a notary to get signed statements. All went as Kelley and the others had expected. Much of the purpose for the trip was being fulfilled in just meeting the pivotal witnesses in the Rountree case, to become familiar with them so they'd be more relaxed when they were called to the witness stand.

The only glitch occurred when they sat down with Charles Tooke. He recounted to Kizer and Ashman the same events he'd told Kelley about, but then he asked the two prosecutors, "I wish you could tell me about that phone call from the neighbor's yard," he said.

"What phone call?" Ashman asked.

Coby Kelley, Tooke insisted, had told him that Piper had made a call on her cell phone from the neighbor's yard just minutes before the shooting. Kelley would later say that Tooke had misunderstood, while Charles would maintain he knew he was right. The call had, in fact, been a few miles from the Hearthglow house early the morning of the shooting.

From that point on, Tooke's relationship with the prosecutors cooled. In the end, as the trial approached, it would become frigid, and the defense would try to use that to their advantage.

17

I n December, Duncan Reid filed a motion asking for saliva samples from Piper Rountree, swabs taken of her throat using a buccal brush that could be used to break down her genetic material and process her DNA. They had the T-shirt Carol Freed had given them to compare it with, along with items from the rented van and the murder scene. The court granted the request.

Another matter came to light that month that caused concern in Murray Janus's office: Prosecutors discovered that one of Fred Jablin's life insurance policies—$200,000, taken out years earlier through TransAmerica—still listed Piper as the beneficiary. All along the prosecutors had coupled two motives for the murder: Piper's desire to regain her children and the possibility of financial gain. In the past, they'd had a rather circuitous argument to establish the money motive: If Piper had custody of the children, not only would the nearly $900 a month child support payments she'd been ordered to pay Fred end, but with Fred dead, Piper, as the children's guardian, could have expected to control all the family assets—Fred's money plus his substantial life insurance. Now the path was even clearer: With Fred dead, Piper individually stood to inherit a $200,000 bonanza.

Meanwhile, in Houston that December, Mac and Tina broke up. Perhaps it was the pressure of the impending trial. Tina would later say Mac had bought "into the Cari stuff." Whatever happened, Mac moved in with Charles for a month, then got his own place. By then Mac was working full-time as a landman on the Galveston project Piper had been researching in the months before the murder. Tina, it would seem, lost no time in recouping her life. In less than a month she'd married again, this time to a Birmingham, Alabama, physician who she said she'd met on Match.com.

"We connected and fell in love," she'd say. "Sometimes, it just happens that way."

In the Henrico jail, Piper was pursuing other interests. She wrote to Jerry Walters, although he didn't write back. In one letter, she claimed she'd taken a mail-order course and become an ordained minister. "For years I've ministered to my own family . . . Sometimes in jail, I help [other inmates] with questions on spiritual matters," she'd later say. "I really try to keep a low profile on that front because I don't want everyone to know the spiritual side of me. I've already had two women ask me if I can marry them."

When she wrote to Charles Tooke, she asked for reading material. He mailed her a Bible. Despite her self-proclaimed spirituality, she wrote back saying, "You can only pray so much," and asked for a copy of *The Da Vinci Code*.

Throughout December and January, Piper wrote letters to the children. She said she loved and missed them and pleaded with them to write her. She recounted the tedious details of jail life, lunches of bologna and potatoes and Tuesday uniform changes, the particulars of her cell and how she was staying fit by running around her room. Somehow, she'd estimated that forty-two laps equaled a one-mile run. The jail was noisy, and she said she slept on the floor. Not having to cook was the only good thing about being in jail, she said, yet she lamented limited coffee and no soy milk, yogurt, or fresh fruit.

Overall, she said, the jail reminded her of a high school slumber party, and she amused herself by cutting out snowflakes with toenail clippers, since scissors weren't allowed.

In her letters, Piper told her children she'd been ill and that it made her cry to think of them without her. She wrote of Jean's wedding, saying they were missed, and asked them to call her mother—their grandmother—and the other Rountree relatives: aunts, uncles, and cousins. And she recommended they find healthy ways to cope with all that had happened in their young lives, suggesting they express their feelings through their artwork.

She even coached them on how to pray and accept Jesus into their lives. God was there, she said, they just needed to reach out to Him for help. He had a plan for all of them, she advised, despite the recent sadness in their lives.

When Callie's birthday approached, she asked what the soon-to-be nine-year-old wanted, then sent a card with handwritten musical Happy Birthday notes that said she wished they were together, followed by X and O kisses and hugs.

At Christmas, she quoted Bible verses and apologized for not working out a way they could all be together. Always, she said she slept with them in her dreams, and signed her letters with Momea and hearts.

What she never mentioned was their dead father.

Jocelyn wrote back in mid-December. Understandably, she said she missed her mother and that she worried about her. She asked if she could send her food and earplugs, so Piper could block out the prison noise and sleep. Throughout the letter, Piper's oldest sounded upbeat and happy, saying she was settling in to her new school, making friends. She was living farther north, and she commented that it was colder there than in Richmond, and said she hoped Piper was warm.

Of them all, Callie, mourning her father, was the one who wrote to tell Piper how sad her life was. The youngster's words bled onto the pages of a Christmas card: *To the Mom I Love Most of All.* She said that her heart was broken, because now she had neither a mother nor a father.

In her response, Piper ignored the little girl's anguish.

Instead, Piper gushed about her joy at receiving the card, and told Callie of her Christmas in jail, with "fake" turkey and all the fixings.

Only Jocelyn would go to the jail to visit her mother. Later, Piper would say the meeting was "joyous," a wonderful event. "You just can't separate us," she said. Yet, Piper also claimed she understood why the others hadn't come, blaming it all on her dead ex-husband. Fred, Piper said, had used her to control the children. Despite her having spent a long weekend with them just weeks before the murder, she claimed: "Fred told the children, 'If you don't behave, you won't see your mother.' " She then launched into a legal argument she said proved she should be allowed to be with them: "The Constitution guarantees me my children, and it guarantees my children their mother."

As 2005 began, Piper wrote the children again. She'd been transferred to another jail, where she said the food was better and fresher, and she was allowed outside during the day. For the past four years, ever since the divorce, the children had had one new address after another for their mother. Now her official address was the Henrico jail off the New Kent Highway.

Meanwhile, Murray Janus and Taylor Stone were receiving large brown envelopes from the Henrico Commonwealth's Attorney's Office filled with discovery evidence against their client. As it began to pile up, the extent of the evidence must have been disheartening: flight records, cell phone records, ATM charges, the wigs, the hotel and car rental records. "They had a shitload of evidence," says Stone, who went to see Piper at the Henrico jail. She still insisted all the records, even the videotape, were wrong. She'd been in Houston that weekend.

In one folder the prosecutors delivered to Janus in mid-December there were thirty-six different tabs detailing different pieces of the prosecutors' evidence.

As the defense attorneys combed through it, Stone and Janus surmised that only two records appeared to help them. One was the packet of records from Houston's Hobby Airport. They showed that the Jeep with Piper's license plate checked in at 5:00 A.M. that morning. That didn't jibe with another record that showed Piper had made a deposit at her bank in Kingwood at 9:25 that same morning. In a trial, all Janus and Stone needed to clear Piper was reasonable doubt. Enough such discrepancies and maybe they would have it.

To collect more evidence to help their client, in January the defense team followed the prosecutors' route and flew to Houston. It would turn out to be a grueling trip. Charles Tooke agreed to be their chauffeur, driving them to talk to witnesses. Beforehand, he'd told the prosecutors of his intent. Piper was his friend, but he didn't want to favor one side over the other.

Tooke first drove Janus and Stone to Piper's Kingwood house, by then locked and deserted. They knocked on doors to talk to neighbors but found no one who saw Piper at the house that weekend.

At Tina's house, Tooke stayed outside in the car while Janus and Stone went in to meet the woman whose name was all over the case, on airplane, rental car, and hotel records. Inside, Tina ranted about Fred and claimed Piper was innocent, yet she told them in no uncertain terms that she would not put herself in further jeopardy by testifying at her sister's trial. If they called her, she insisted she would take the Fifth Amendment: refusing to testify on the grounds that it could incriminate her.

After an hour and a half the two defense attorneys left the

house. Tooke thought they looked green from the experience.

Since it was a sunny, warm day in Houston, Janus and Stone walked, while Tooke followed in the car, to Under the Volcano. There, they talked to the owner, bartender, and a few customers, asking if anyone had any knowledge that could help their client. Did anyone remember seeing her there the Friday evening before the murder? Had they heard of anyone who remembered seeing their client in the bar that night? Janus and Stone again came away disappointed. No one came forward to shore up Piper Rountree's crumbling alibi.

From the Volcano, they went to McVey's office, where they talked to the boisterous attorney about Piper, Tina, and what he knew about the case. They asked to see the .38 Piper had given him years earlier, the one Mac had asked her to dispose of. McVey obliged, pulling it out of a chest of drawers.

"I know it wasn't this .38," he said. "This one has never left this drawer."

An hour or so after they arrived at McVey's, Tooke saw Tina pull up in her car. She went up to him and started ranting, talking about Mac, detailing aspects of their love life and how she'd had other lovers during their time together. She spouted conspiracy theories about who had killed Fred Jablin, including one Piper had been promoting, that someone at UR was behind the murder.

As they walked out of McVey's front door, Stone and Janus saw Tina and headed directly for the car. Neither looked pleased. When McVey came out, Tina cornered him, ranting about things Charles couldn't make out. While Tina was busy with McVey, Charles hurriedly drove off with Janus and Stone in the car.

As their time in Houston wore on, it seemed the defense attorneys would get no good news.

When Janus asked Charles about the e-mail and call he'd

received from Piper the afternoon of the murder, Charles told him, "Of course she could have sent that from anywhere, as long as she had access to the Internet."

"Is that true?" Janus asked Stone.

"It is," Stone said. "You can even send e-mails from a cell phone these days."

When Taylor Stone called Cheryl Crider, the bartender, she refused to talk to them. Kevin O'Keefe was more obliging, but when he finished answering their questions, he said to Janus, "Doesn't seem like you have a lot that points to her innocence."

While many of those who talked with the defense attorneys told them that they didn't believe Piper was capable of murder, none offered any proof that could be beneficial in the upcoming courtroom battle.

Two days after they arrived in Houston, Janus and Stone flew back to Richmond.

As the days progressed and the trial neared, Janus appeared to grow even more apprehensive about the case. At one point, when Wade Kizer approached him about the trial, the aging attorney looked at the prosecutor and shook his head. "Wade, I hope one day you have a lawyer for a client," Janus said.

Kizer chuckled softly.

Still, he understood where Janus was coming from. By maintaining her innocence in the face of so much evidence, Piper Rountree had tied Janus's hands. If she admitted she'd killed Jablin, her attorneys could have mounted a defense based on the police reports at the house, claiming she'd been an abused spouse. Or Janus could have used her substantial psychiatric records to her advantage in a not-guilty-by-reason-of-insanity plea. If it didn't earn her an acquittal, it could sway a jury to recommend a light sentence. When Piper insisted she was innocent, that she was in Houston at the time of the murder, neither argument was an alternative.

Her pleas of innocence also blocked any possibility of a plea bargain, although Janus doubted that was in the offing. He knew, with all the evidence against her, with Fred Jablin's stature in the city, that Kizer would be criticized if he offered Piper Rountree a plea bargain. One day he said as much to Kizer.

"I know there's nothing you can offer that she'd take," Janus said.

Kizer answered, "Murray, you're right."

Then, just when all looked darkest for the defense, something happened.

The hearing began on January 28, 2005, three weeks before the beginning of the trial. At issue were motions Murray Janus had filed on behalf of his client, Piper Rountree. He had four requests. The first two were a reduction in Piper's bond, so she could get out of jail until the trial, and a change of venue, moving the trial to another county, due to the heavy local media coverage of the case. A feature article had run in the *Richmond Times-Dispatch* just weeks earlier by reporter Paige Akin, one in which she recounted the contentious divorce and interviewed people out of Piper's and Fred's pasts, along with a handful of the potential witnesses.

Janus had also filed motions to suppress the evidence supplied by Carol Freed and all the witnesses who'd identified Piper as flying under Tina's name and being in Virginia that weekend. In his briefs, the defense attorney argued that the identifications were overly suggestive, since police had shown potential witnesses not a spread of photos but only one, a photo of their main suspect, Piper Rountree.

While not overly confident, the prosecutors, Kizer, Ashman, and Reid, weren't particularly worried about the hearing. They believed Virginia law was on their side, including in the case of the photo identifications. Reid had written his own brief for the judge, one that cited state law that sanctioned the use of one-photo identifications under certain conditions—if the IDs were soon after the sighting; and the witnesses were not emotionally distressed, with ample time to observe the subject—all factors that fit the Rountree case.

Yet even before the hearing began, Kelley was worried. He'd seen Piper in the jail, and she'd put on weight, nearly twenty pounds in the past three months. "I had to take a double take. I barely recognized her," he told Kizer. "She's trying to mess with our witness IDs."

It would turn out that some of the witnesses would have as much trouble recognizing Piper Rountree as Kelley did.

That day in the courtroom, the three prosecutors sat at a table on the left of the courtroom, and Janus, Stone, and Piper Rountree sat at a table on the right. The room was formed in a semicircle, with all the chairs pointed toward the witness stand and the judge. Although the jury box was empty, the room was crowded, filled with reporters from the local newspaper, the *Times-Dispatch*, and local and national television stations.

The first witness on the stand that morning was Kathy Molley, the Southwest Airlines ticket clerk. As Ashman began the questioning, she felt confident. Molley had talked about Piper's luggage, her wig, the gun, even describing the shoes Piper had in her luggage. She'd been able to depict the way Piper looked and acted in detail, and had shown no hesitation when Kelley showed her the photo. But when Owen asked Molley if the woman she'd seen at the airport was in the courtroom, the ticket agent grew quiet.

At the defense table, dressed in a dark blue, double-breasted suit, Piper sat taking notes, head down, looking every bit the attorney. When Molley had seen her, Piper was not only substantially thinner, but wearing a blond wig; now her hair was her natural color, a dark brown.

"I don't see her," Molley said.

Ashman sucked in her breath, feeling a sense of panic.

Molley was removed from the courtroom, Ashman nervously took her seat, and Duncan Reid and Murray Janus approached the judge.

Before the judge, Reid contended the in-court and out-of-court identifications were two different matters, and that while Molley hadn't been able to pick out Piper Rountree in court, her identification of the photograph should be admissible during the trial. "There are a number of Virginia cases where out-of-court identifications are admissible," Reid said.

"The lady looks around, as she did, the entire courtroom and says, 'No. I did not see the person I'm talking about,'" Janus countered.

As the judge read over the prosecution's brief, Reid cited cases. He agreed the police had only used one photo, yet he argued, "The identification is nevertheless so reliable that there is no substantial likelihood of misidentification." Throughout the argument, he continually reminded the judge that Molley's identification of Piper from the photo was within days of the murder, not three months later, as the pretrial hearing was.

After the arguments, the judge took the matter under advisement, and the ticket agent again took the stand.

To increase the chances that the judge would allow Molley's identification of the photo, Ashman asked her, "How long did you talk to investigators before being shown a photograph?"

"Forty-five minutes," Molley said.

"Okay, and when you were shown a photo, what was your reaction?"

"Right away, that's her," Molley said.

When Janus took over on cross exam, he pursued his theory that Kelley had presented the photo in an overly suggestive manner, asking, "Do you recall what Mr. Kelley said to you, 'We have a picture here that we think is the person who purchased the ticket'?"

"He asked me if this is the woman that you just described," she said.

"But he did not tell you that we think this is the woman that you described?" Janus countered.

"No, sir," she said.

"Do you recall that this woman paid cash?" Janus asked.

"No," Molley corrected. "She paid with a credit card."

Molley's testimony wouldn't be the end of the prosecutor's problems that day in the courtroom. Boyd Adams, the man who'd checked Mac and Piper in at the 59 Gun Range, picked her out easily at the defense table. But Ray Seward, the Eagle rental car owner who'd checked the rented van in, couldn't find Piper when Owen asked him to, despite having said he was a hundred percent certain he'd seen the woman in the photo. The same thing happened when Tina Landrum—the Miller Mart assistant manager who said Piper had alcohol on her breath the morning of the murder—looked around the courtroom. Landrum wasn't able to point to the woman in the photo.

The prosecutor's hopes looked dashed again when Tarra Watford, the Eagle car rental clerk, took the stand.

"Do you see the woman who came into the dealership that day?" Owen asked.

"No," Watford said.

"Keep looking," Owen instructed.

Watford searched the room again, this time stopping and pointing at Piper. "It looks like her," she said. "But her hair is dark."

"Let it be known that she identified the defendant," Owen said.

Then Piper said something, and Watford shouted, "That's her."

"Are you sure?" Ashman asked.

"Yes. When I heard her voice, it's distinctive."

With that, Piper looked up and stared at Watford. While

she may have intended it in a threatening way, Watford took the opportunity to look closer and came away more certain Piper was the woman to whom she'd rented the maroon van.

Still, even those who identified Piper often offered testimony fraught with discrepancies. Allan Benestante, the airport security agent, under questioning by Wade Kizer, was able to point to Piper in the courtroom, but he described the woman he saw as "up to my forehead."

At five-eleven, Benestante towered over the petite Piper.

Kizer asked, "Do you know what kind of heels the passenger had on?"

"I know they weren't sneakers," the man with the deep circles around his eyes said.

When it was over, three of the seven witnesses couldn't pick Piper out in the courtroom. On the stand, Murray Janus cross-examined Coby Kelley about the way he'd obtained the identifications.

"Did you say anything to Mr. Benestante, such as 'This is my suspect,' or 'I think this person committed the murder,' or any leading-type statements like that when you showed him the picture?" Janus asked.

"No, sir," Kelley answered.

"Was there any hesitation in Ms. Molley's voice when she identified the woman in the photo as the one she'd checked in?" Wade Kizer asked.

"No," Kelley said.

"Did you ever say, 'I'm investigating a murder in Virginia and this person is the person I think did it,' or any words to that effect?"

"No, sir," Kelley said.

In the end, the prosecution won on every point. Piper would remain in jail and the trial would continue in Richmond, unless it became evident that an impartial jury could not be seated, and the prosecutors were able to put on evidence that the witnesses who couldn't identify Piper in the courtroom had done so in a photo soon after the killing.

Another point Janus had argued was that the tape-recorded interviews Kelley had made of Piper were coerced. Here, too, the judge disagreed, saying they would be admissible during the trial.

The defense, however, had won an advantage: They would be able to tell the jurors that at a pretrial hearing, three of the eyewitnesses didn't recognize Piper Rountree. And they'd be able to point out discrepancies in the other witnesses' descriptions of the woman they'd seen, many of which more closely resembled Tina than Piper. That helped lay the foundation for the alternative Murray Janus would offer in the courtroom: that rather than Piper, Tina may have committed the murder.

With that matter settled, one more piece of evidence remained fluid. The only evidence Virginia law decrees a defense attorney must share with a prosecutor is the accused's alibi. As the case progressed, Piper Rountree's version of her whereabouts the weekend of the murder continually changed. At first she claimed she'd been home, at the house in Kingwood throughout that Saturday, the day of the murder. By mid-February she was saying something markedly different, that at three or three-thirty that Saturday afternoon, when the plane was still an hour away from landing at Hobby Airport, she'd been in Marty McVey's office.

As the trial approached, Duncan Reid sent Murray Janus information on the K-9 unit that had gone to the murder scene, to inform him that the dogs had not tracked by scent but "ground disturbance," and that they had walked in front of the McArdles' house, where the footprints were seen that day, then turned right at the corner and right on Helmsdale, the block that backed up to Hearthglow. The dogs had finally stopped at a manhole cover, which was pried open. Nothing was found inside but a decomposing animal.

Meanwhile, from jail, Piper was still writing the children,

now telling them that angels were guarding them. She claimed her only fault was that she was too trusting in people, and theorized that hell was not in the afterlife but in the suffering experienced on earth. She seemed confident, promising them that they'd be reunited and all would end well.

At about that time from Houston, Charles Tooke was still pushing the prosecutors to tell him about the phone call he said Kelley had described as originating in the neighbors' yard the morning of the killing. In an e-mail, Owen Ashman told him, "It didn't occur that way."

As the trial date neared, more issues cropped up with Tooke's testimony, with Charles insisting some of what he'd told Kelley about Piper was shaded in the officer's written account to make her sound guilty. Concerned about playing fair, when Tooke e-mailed prosecutors to complain, he also sent a copy of the letter to the defense.

The week of the trial, Duncan Reid followed up on something Wade Kizer had noticed when looking at the checks cashed out of the Jerry Walters Wells Fargo account. At 9:55 on the morning Piper left Houston for Richmond, a check was written for $21.64 at an Academy Sports and Outdoors store near her home in Kingwood. When Reid received a detail of the purchase, another piece of evidence was slipped into the growing evidence file: The receipt was for a Masterlock cable-type, combination lock, the same type of lock Allan Benestante and Kathy Molley would testify they saw wrapped around the gun Piper had checked on the airplane.

"Bingo!" Reid said, grinning.

Meanwhile, Owen Ashman, without telling her colleagues, was doing a little investigating of her own. She and the other prosecutors worried about the Hobby Airport parking records that showed Piper's Jeep was in the garage at 5:00 A.M. on Thursday, October 28. The flight to Baltimore they contended Piper was on didn't take off until eleven-twenty that morning. Now they also had the Academy check for the cable lock, which suggested she was in Kingwood at nearly ten that morning, nowhere near the airport. Their Hobby parking lot witness, John Guidry, theorized that all cars that entered the lot on a given day were logged in airport records as having been in the parking structure at 5:00 A.M., but that seemed odd to Owen, and when questioned, Guidry appeared uncertain. On the stand, Kizer feared he'd be a poor witness.

To keep from confusing the jury, Kizer was considering not putting Guidry on the stand and foregoing entering the parking lot records into evidence. But it was damaging to the defense, and Ashman wanted to use it. So she took a chance. Kizer was a careful man who didn't like to open doors unless he felt certain he knew what was on the other side, and Owen knew he wouldn't approve. So without telling him, she called Breck McDaniel and asked them to plant a car at Hobby Airport late one morning.

When the parking lot records on the planted car came back, Ashman was elated: The second car showed up on the records just as Piper's Jeep had, as having been in the garage at 5:00 A.M. After she went over the paperwork on the second car with Guidry, he was more confident, and Kizer, after reviewing the new evidence, agreed: He had no reason to keep Guidry and the parking lot evidence from the jury.

As the days counted down to the trial, the evidence continued to mount.

Over the telephone one afternoon, Jerry Walters mentioned something to Kizer that got the prosecutor's attention. Walters hadn't been writing Piper, but she'd written him off and on from the jail. Her latest letter included a rather odd suggestion. "She asked me to marry her, so I wouldn't have to testify," Walters said.

"Would you send me the letter?" Kizer asked. Walters agreed.

When Walters's letter from Piper arrived, Kizer read: " 'I wonder about the "us" in all of this . . . I hope and pray you will write me your thoughts . . . I know you always have plan A, B, and C . . . do you mind sharing them with me?' "

It was what Piper had written deeper in the letter that Kizer found fascinating: "I have wondered again and again how to spare you all of this. It has occurred to me that perhaps they can spare you further or any testimony if we are husband and wife?" From there, Piper explained that spouses couldn't testify against one another and maintained that she and Walters could reasonably argue that, because of their relationship, they were common-law married. "I have to tell you that you *really*, *really*, *really* don't want to testify if you don't have to," she wrote.

Much of the rest of the letter was classic Piper, in which she saw the world only as it affected her, including her rules for how a man should treat her. Jerry, she said, was to always be supportive and there when needed. "Always tell me you love me . . . always respect, loyal, wear a Superman cape . . . come to my rescue and protect me from evil and badness, always trust, hope, and have faith, no matter what." She even sent along a booklet, one entitled: *Building Blocks to a Strong Marriage.*

To Kizer, the letter with Piper's proposal was another strong indication of her guilt. Why would she be worried about Jerry Walters's testimony unless she knew what he had to say could hurt her?

As might be expected, all these last minute bits of evidence weren't good news for the defense. When the prosecutors' disclosures arrived, they delivered one disappointment after another. Owen Ashman's gamble on the Hobby records dashed Janus's hopes of using Piper's deposit at the bank around nine that morning to contradict the parking lot records. The information from Academy Sports about the cable lock was worse news, another link in the chain that tied

Piper to the .38 caliber gun and the plane ride to Houston. And then there was Jerry Walters's letter, which Kizer included a copy of for Janus's consideration.

It could be argued that at every turn, Janus was being sabotaged by his own client.

In the days leading up to the trial, both sides assessed their cases' strengths and weaknesses. Wade Kizer saw the evidence against Piper Rountree as powerful. Yet, he knew it was ultimately a circumstantial case. Kizer had no eyewitnesses to the murder and no DNA or fingerprints that tied Piper to the murder scene. Evidence from the hotel room and the rented van had all come back negative of any trace of her DNA or fingerprints on the forensic reports. The T-shirt Carol Freed had given them tested positive for Piper's DNA, but bore no sign of gunshot residue, Fred's blood, or anything that tied it to the murder. The .38 bullet Jerry Walters had given Kelley didn't match the ballistics of the one that killed Fred Jablin. And the only blood on the scene was the victim's. Finally, Piper had made no incriminating statements or confessions, and police had never found the murder weapon.

As the trial date approached, Owen Ashman played the case over and over again in her mind, in bed at night, in the shower in the morning. The prosecutors had heard from Janus that Tina would not be called to testify, but Owen wasn't certain. What if Tina took the stand and, without admitting any guilt, implied she *could* have committed the murder? What if Tina said just enough to inject reasonable doubt?

In the beginning, Owen had believed that Tina Rountree could have been involved in the planning. But as they investigated, the police had found no evidence that even suggested Tina had been involved with the actual murder. Nothing beyond perhaps helping her sister dispose of evidence after the fact, the offense she was charged with in Texas. "We'd had lots of suspicions, but that's all they were, just suspicions," says Ashman. "The bottom line was that everything we had pointed to Piper committing the murder alone."

Across town in their cluttered Franklin Street offices, Janus's vast collection of owl figurines staring at them through saucer-wide eyes, Murray Janus and Taylor Stone planned strategy and readied their witnesses for trial. They saw the Rountree case from an entirely different angle. "Tina Rountree was the name that repeatedly kept popping up," said Janus. "Many of the witnesses described a woman who was taller, heavier, with blond hair. That description fit Tina, not Piper."

The defense strategy was simple: plant reasonable doubt by continually reminding the jury that there were two Rountree sisters, and suggest that the guilty one might not be the sister sitting before them in the courtroom.

· ·

18

Of course, I wanted to be with Piper, to support her during the trial," Tina said thoughtfully. Then she frowned, perhaps recalling all they'd been through together, two sisters who were integral parts of each other's lives from childhood on. Yet Piper described Tina as more than her sister; rather, a surrogate mother. With eight years between them, perhaps Tina saw her little sister in the reverse, almost as one of her children. "Piper and I were *always* there for one another," she whispered. "I wanted to be by her side, but I just couldn't. I had my own issues to consider."

While Tina remained in Houston facing evidence tampering charges, Richmond readied for Piper Rountree's trial. Jury selection began on the morning of February 22, 2005. Many on the jury panel said they'd heard of the case, as Murray Janus had predicted, but most insisted they could put what they'd heard aside to listen to the evidence, and within a few hours a jury of seven men and seven women was seated, including two alternates.

At two that afternoon Wade Kizer rose to give his opening statement. In the audience, Michael Jablin sat behind the prosecutors, guarded from the media by a Henrico County victims' advocate. On the front table on the right side of the curved courtroom, Janus and Stone sat with Piper. Behind them, often clutching a tissue and dabbing at tears, was

Piper's mother, Betty Rountree, and her oldest son, Piper's brother Bill, a Harlingen, Texas, attorney.

For much of the trial Piper, dressed in a severe, double-breasted, navy blue business suit, struck a professional pose, taking notes like the lawyer she was, on a yellow legal pad she slid back and forth with Stone. At other times, she was the mail-order minister, her head bowed and her hands folded before her face, as if in somber prayer. Occasionally, she sobbed, her face, pale from the months in jail, flushed and twisted in pain, especially when her children were mentioned. Throughout the trial, when anyone uttered their names, Piper became visibly rigid.

On the bench, Henrico County Circuit Court Judge L. A. Harris, an avuncular middle-aged man with an ample double chin and a dour expression, peered down at the crowd gathered before him. The red power lights on the television cameras shone steady, indicating they were recording, and reporters scribbled quick notes in steno pads as Kizer addressed the jurors, thanking them for their service and explaining his role and the roles of his fellow prosecutors along with those of Janus and Stone.

"I would like you to have a road map of what we expect the evidence is going to show," he said. Accompanied by a PowerPoint presentation, Kizer, in a matter-of-fact manner, his halting voice booming through the courtroom, systematically laid out the commonwealth's case, beginning by introducing each of the main characters in the drama that would unfold before the jury: Fred Jablin, the victim; Piper Rountree, his wife and the accused; Jerry, Piper's oilman lover; Tina, her devoted sister; Mac, Tina's live-in boyfriend; and all the evidence, from the debit card receipts to the wigs and the cell phone records. What were Piper's motives for the murder? As expected, Kizer pegged them as twofold: reclaiming her children and getting her ex-husband's money.

Piper Rountree had left so many "crumbs" behind, as Captain Stem had referred to the evidence, that it filled Kizer's allotted hour. When the prosecutor projected a photo of her ex-husband's dead body, Fred's shirt cut away by paramedics in a failed attempt to revive him, Piper lowered her head, not looking at the screen.

In conclusion, Kizer asked the jury to "give both sides a fair trial. Piper Rountree is entitled to a fair trial, and the opposite side, the commonwealth, is equally entitled to a fair trial."

When it came to tangible evidence, Janus had remarkably less to work with, yet, as a counterpoint to Kizer's calm deliberate manner, he delivered his opening with an emotional fervor.

Defense attorneys have a twofold mantra: First, that the state or, in the case of Virginia, the commonwealth, bears the burden of proof. Second, perhaps the jury's most important duty: They must presume the woman across the room from them, the defendant, is innocent until proven guilty.

Listen to the evidence, the grandfatherly Janus said: "Does one person say, 'We saw Piper Rountree not a fleeting shadow running away?' . . . Identification is key." No one, he said, could point to his client and say, "We saw her pull the trigger." None of the evidence in the case tied Piper forensically to the murder scene, much less the murder. All the prosecutors had, he held, were shreds of circumstantial evidence, not enough to convict and lock a woman away in prison.

Janus pointed out that Piper had made a child support payment to Fred in early October, weeks before the murder. "She wouldn't have paid child support if she'd been intending to kill her ex-husband," he said.

With all the vigor he was known for, the defense attorney chipped away at the evidence Kizer had stacked before the jury. The wigs were nothing. Tina had many wigs she lent to her cancer patients who lost their hair, he said, and the ones in question, Piper purchased for an innocuous reason: simply to wear on Halloween. The gun range? Piper had gone at Mac's insistence. He was a "gun freak," and "Texans shoot guns all the time."

What about those who said they could identify Piper, like Kathy Molley, the Southwest Airlines ticket agent? Don't be surprised if Molley can't identify Piper in the courtroom, Janus said, pointing out that Molley had failed to recognize her at the pretrial hearing. In fact, Molley's description of the blond woman she saw fit Tina better than Piper.

As to the cell phone calls, Piper and Tina had many cell phones, he argued, and they were used interchangeably by both sisters. The airline tickets, car rental, and hotel? They weren't in Piper's name but in the name of Tina Rountree. "One name you'll hear over and over and over again: *Tina Rountree*," he said with conviction. "The name that will keep popping up is *Tina Rountree*."

To tear down the prosecutor's case, Janus pointed to mistakes in Kizer's opening, including the wrong date for Piper and Fred's marriage, the misspelling of Jerry Walters's last name as Walker, and a zero missing from the address of the Homestead Suites in an exhibit.

"In the end," Janus concluded, "you're going to find there are too many discrepancies."

There were things Janus didn't want said at the trial. In murder cases, the victim is often vilified, his or her character dismantled by the defense, to suggest that even if the jury believes the person on trial pulled the trigger, their guilt is of little consequence, for the victim wasn't worth society's retribution. The murder had, in some way, been justified.

It was obvious to everyone who talked with her that Piper was eager to paint Fred in such a bad light, to say he wasn't the devoted father and dedicated husband he'd been portrayed in the media. She charged he'd been psychologically, emotionally, even physically abusive, not only to her but to the children. To do so, however, could hurt her case, by supplying prosecutors with further motives for murder. Perhaps Piper felt so manhandled by her ex-husband in the divorce, they'd argue, she'd murdered to exact revenge.

"I saw what appeared to be someone running right in front of my house toward the corner," Bob McArdle, Fred's next-door neighbor, told the jurors, beginning the prosecution's presentation of their case. Including McArdle, prosecutors used the first witnesses who took the stand to set the scene the morning of the murder: the shots in the darkness, the search for the source, and the discovery of Fred Jablin's cold, dead body. With each witness, the prosecutors supplied the jury with nuggets of evidence they'd later link into a chain they hoped would support their case: The person jogged; Piper was a jogger. No cartridges were found, suggesting the gun was a revolver. Piper used a .38 revolver at the gun range, and witnesses would testify that she'd carried one onto the airplane.

During cross examinations, wherever he could, Murray Janus whittled away at the prosecutors' evidence. "Was there any attempt made to get fingerprints off the bullets?" Janus asked the forensic officer, Danny Jamison.

"No," Jamison answered.

When Kizer questioned Jamison on redirect, he countered, "Can you get fingerprints off a bullet that has traveled through somebody's body?"

"No, sir," Jamison answered.

Standing before the jury, an assistant Henrico medical examiner, Dr. Deborah Kay, a plain woman with bouffant blond hair, demonstrated the paths the two bullets had taken through Fred Jablin's body. The one through the lower back, she explained, exacted severe damage to his internal organs. The result: Fred had lost nearly half his blood into his abdominal cavity. The damage, Ann Davis, the commonwealth's ballistic expert, testified, wasn't unexpected, as the type of bullet used, a .38 special, was designed with a hollow point that mushroomed inside the body, making it exceptionally lethal.

"It does more damage?" Kizer asked.

"Yes, sir," Davis answered.

After the murder scene and Fred's death had been explored, the prosecutors turned their attention to the second phase of their plan: proving that Piper Rountree was not in Texas, as she'd claimed, but in Virginia on the day of the murder.

The first such witness to take the stand was the airline ticket agent, Kathy Molley. As she had at the pretrial, Owen Ashman led Molley through her encounter with the blond woman with the cute shoes who checked a .38 revolver on the Southwest Airlines flights that brought her from Houston to Virginia. Nothing in Molley's testimony was truly at issue, as far as the defense was concerned, except for one point: the identity of the woman with the gun.

Molley explained how, when shown a photo of Piper Rountree, she had identified her as the woman at the airport.

Then Ashman held her breath and asked, "Do you see that woman in the courtroom today?"

As at the prior hearing, Molley quietly surveyed the courtroom. This time, however, the result was different. "That lady looks familiar, but her hair didn't look like that at all," she said, pointing at Piper.

"Would it help if you got closer?" Ashman asked. Molley said it would, and she got up and walked toward the defense table, standing right before Piper Rountree, who stared down at the legal pad on the table.

"Is that the woman you dealt with?" Ashman asked.

"Yes," she said.

"Let the record reflect she identified the defendant, Piper Rountree."

At the defense table, Piper looked up, shaking her head an exaggerated no.

When he took over, Janus pointed out that Molley, who'd returned to the witness box, hadn't been able to identify Piper at the pretrial hearing. Molley didn't disagree. It was true, and Janus had made his point.

Yet on redirect, when Ashman took over, she asked Molley why the result had been different on this day than at the previous hearing.

"I saw her closer today," Molley said. ". . . It's the shape of her face."

With that, the judge adjourned the courtroom, ending the first day of trial.

In Houston that evening, Team Martini met at the Volcano. Many had friendly wagers on the outcome of the trial, their usual bet, a martini. Each day, they'd check the *Richmond Times-Dispatch* website for the newspaper's updates on how the trial was progressing. Kevin O'Keefe and the bartender, Cheryl Crider, were both in Virginia waiting to testify for the prosecution. No one who'd seen Piper when she came in to the Volcano to ask if they recognized her agreed with Murray Janus's description of the event in his opening statement: that Piper wasn't trying to set up an alibi but simply looking for someone who remembered her from that Friday night. "The lady came in here with a notary, into a bar with a notary, and she was peeved when Kevin and Cheryl refused to sign a statement," says Ace, a tall, middle-aged business exec leaning on the bar while swirling his martini under a hanging tin light. "Sure looked like she was alibi shopping to all of us."

At the University of Richmond, news reports of the trial played in dorm rooms and gathering areas throughout the campus. "Many of us felt there had to be justice," says one of Fred's students. "We felt like his death needed to be accounted for."

When Piper visited with Loni at the jail, she insisted she had no reason to worry. "Piper seemed certain she'd be acquitted," Loni says, running her fingers through her thick blond-highlighted hair. "I never said it, but I thought, 'Peeps, are you looking at the same evidence I am?'"

At the trial each day, Michael Jablin, a tall, dark-haired, balding man with glasses, kept a photo of his murdered brother in the pocket over his heart. At times he took it out during testimony, ran his thumb over it, and his eyes welled with tears. When he glanced at Piper, she quickly looked away. He must have remembered the happier times, when Fred was head over heels in love with the artistic attorney who could be the life of the party.

"I think she looked ashamed to look at me," Michael Jablin would say later. "It was hard for me to look at her."

The second day of trial began late the next day, Wednesday, allowing the attorneys to attend the morning funeral of Judge Robert Merhige Jr, who'd died the previous Friday. For Murray Janus, it was a personal loss. Merhige had been his friend and mentor, the man who'd taught him what it meant to be an attorney.

As Judge Harris called the court to order that afternoon, the prosecutors' witnesses congregated in a secluded back room, waiting for their turns on the witness stand. Many of them had flown in from Houston and would remain in Richmond throughout the trial. Jerry Walters hadn't wanted to come, but finally relented. Charles Tooke, Dean Lowry, Mac McClennahan, and Carol Freed were all standing by, while behind the scenes the prosecutors were considering their lineup. They had two concerns: Charles Tooke had been critical of the police, questioning statements Coby Kelley had made to him, and they weren't sure they wanted to put him on the stand when Janus could use that to Piper's advantage. And Carol Freed? Her testimony could be powerful, but did they trust her to be a good witness? None of them had forgotten how she'd behaved when they were in Houston. It didn't help that in the witness waiting room the defensive driving comic cracked jokes, entertaining the other witnesses. At one point she suggested, laughing, that they all testify that they'd slept with Piper.

Some found her funny, but others were not amused.

"Cari was in-your-face," says one who was there. "Loud and obnoxious."

In the courtroom that afternoon, Susanne Shilling, Fred's attorney, took the stand and described the acrimonious Jablin-Rountree divorce.

"How many times did you go to court?" Ashman asked.

Janus objected, but Judge Harris allowed Shilling to answer, "Many times. Innumerable times."

Into the record, Ashman placed the final divorce decree, and records of Piper's arrears, $9,792.64 in 2003, and, by the time of the murder, still more than $7,000. Ashman also pointed out that Fred had fought Piper's bankruptcy.

"It's not unusual in divorce cases, particularly when custody is involved, to have them contested and get contentious?" Janus asked.

"No, sir, it's not unusual," Shilling answered.

In addition that afternoon, prosecutors offered another bit of financial evidence: Fred's $200,000 life insurance policy that still listed Piper as the beneficiary.

Yet a motive didn't prove Piper Rountree had committed murder. The jurors would need more.

When Allan Benestante took the stand, the airport security officer described the gun he said Piper had taken onto the plane: a small chrome .32 or .38 caliber revolver. After putting the gun in Piper's hands at the airport, Duncan Reid called a witness from Piper's cell phone company, Sprint, to the stand, to place her firmly in Virginia.

Through Reid's questioning, Crystalee Danko, a Sprint

records official, explained how cell phones worked, bouncing off towers, their locations recorded at the company's main office. Piper's cell phone ending in 7878 had been used in Richmond the Thursday before through the Saturday of the murder, Danko testified. There was no doubt about it.

With that, Duncan Reid entered commonwealth's exhibit number twenty into evidence, a summary of the cell phone records of Piper Rountree.

Along with the whereabouts of Piper's phone during the days preceding and the day of the murder, Reid emphasized one other insight given by the records: Piper was a phone-aholic who commonly made dozens of cell phone calls daily; yet throughout the three days she'd spent with the children in early October, she hadn't made a single call. At this point in the trial it seemed an unimportant detail, but Reid had a reason for wanting that spelled out before the jury, one that he and Kizer would reveal at the end of the trial.

The whereabouts of Piper's cell phone established, Reid then carried the testimony one step further, putting on the stand witnesses who could testify that the person who'd made those calls on the 7878 number was Piper.

The first was Steve Byrum, whom Piper had called twice on October 28, two days before the murder. Both were calls Sprint recorded as being made in the Richmond area. Whose voice was it on his telephone voice mail? Reid asked.

"Piper's," Byrum replied.

The prosecutor had made his point: Byrum said Piper had made the calls. Sprint showed that the calls were in Richmond. Therefore, the inescapable conclusion was that at the time the calls were made, Piper Rountree was in Richmond. As he'd do with each of the coming witnesses, Reid then had Byrum write his name next to his phone number on the exhibit, labeling his calls from Piper.

On cross exam, Janus asked Byrum, who'd had a brief

relationship with Tina during a trip to Houston, if she and Piper had similar voices. "They're similar but I knew them well enough that I could tell the difference," Byrum said.

"Do you recall asking Piper on the telephone if she was in Richmond on the day of the murder?" Janus asked.

"I may have," Byrum answered.

"Do you recall her telling you she was not?" Janus asked.

Kizer objected. Judge Harris agreed and sustained the objection, but Janus had skillfully slipped into the testimony his client's denial that she'd been in Richmond without putting her on the stand.

After Byrum, Dean Lowry testified. Yes, Piper had called him on the twenty-ninth, the day before the murder, he told Kizer, and he'd had no doubt she was the one on the telephone. Lowry also testified that in the days following the murder, Piper had told him she'd thrown her cell phone away in pieces, and that rather than run to Virginia after Fred's death to comfort her children, Piper hid out at a Houston hotel, expecting to be arrested.

The Jablin children had been unseen but not forgotten throughout the trial. Susanne Shilling had testified how aggressively both parents had fought for their custody. When they entered the courtroom that second day, it was not a physical presence but through two statements read to the jury.

The decision had been made by the defense before trial: Rather than force Paxton and Callie to take the stand to testify that they had talked with their mother on her cell phone the day before the murder, at a time when records showed the signal was bouncing off towers in Virginia, Piper would stipulate their testimony. That meant Owen Ashman could simply read their statements into the court record and to the jury. Yes, Paxton and Callie had talked with their mother that Friday. She'd told them both she was in Galveston on her way home from work. And, yes, their father kept to a schedule. Fred Jablin routinely left the house between six-fifteen and six-thirty, to retrieve the newspaper, something his wife of nearly two decades would surely have known.

Not forcing her own children to testify against her was, perhaps, the first truly kind thing Piper had done for them in a very long time.

Another youngster did take the stand that afternoon: Russell Bootwright, Paxton's good friend, who testified about the day before the murder when he and Paxton were playing poker with friends in a garage. Paxton's cell phone rang, and the youngster had appeared thrilled to hear from his mother. To get some privacy, Paxton took the telephone and walked away from the others. As Russell testified from the witness stand, Piper smiled at him.

"Nickel and dime stuff, I hope?" Reid asked, regarding the poker.

"Yes, sir," Russell replied nervously.

After she'd talked with Paxton, Piper had asked to talk with thirteen-year-old Russell, and he had gotten on the phone with her. From the audience, Russell's dad, Bill Bootwright, watched, concerned about the toll the testimony was taking on his son.

"Did you recognize her voice?" Reid asked Russell.

"Yes, sir," the teenager said.

In the hallway outside the courtroom, once Russell's testimony was over, Bill comforted his son, who cried, upset at having to testify against a woman he and his family considered a friend. The Bootwright family had met the Jablins when their boys were in second grade at Pinchbeck Elementary, and they'd all grown fond of Piper, finding her fun and enthusiastic, a model mom. It was hard for them to believe she was capable of murder. He was so concerned that she be treated fairly, that Bill had been in the courtroom since the first witness and planned to stay for the duration, to judge her guilt for himself.

He didn't believe Piper had murdered Fred. None of the family did.

After starting late, the trial continued into the evening to make up for lost time. One of the final witnesses before the court adjourned was a reluctant Jerry Walters. He, too, testified that it had been Piper using her 7878 cell phone in the days before and the day of the murder. For Reid, Walters wrote his name next to his cell phone number on the chart, filling in yet more of the empty spaces.

On the stand, Walters then read in his Louisiana drawl from the letter Piper had sent him from jail, now marked commonwealth's exhibit number thirty-two, the one in which she'd proposed marriage. "I have wondered again and again how to spare you all of this," Piper had written. ". . . By law, a husband could not testify against a wife, and vice versa."

After the murder, Walters recounted how he'd checked the Wells Fargo Bank account in his name and discovered it had been used in Richmond the day before the murder. Soon the bank records would be in evidence, detailing what the debit card had been used to buy, including the wigs and latex gloves, size small.

On the stand, the tanned, broad-shouldered Walters put on a pair of reading glasses to examine the paperwork for the Mail & More, P.O. box number 162. Yes, he said, it was Piper's, and she'd added his name to the box.

The more evidence the prosecutors put before the jury, the more those who knew Piper wondered. Dean Lowry, for one, felt certain that Piper had a loophole to fall back on. "She was too smart," he said later. "I couldn't believe she'd leave all that evidence out there, without a plan to explain it away."

Meanwhile in Austin, Fred's old assistant, Lora Maldonado, considered the same thing, how a woman with Piper's intellect could have been so careless. At first it troubled her, then, eventually, she thought back to the Piper she knew, and it seemed to make sense. "In a way, it was just like Piper," she'd say. "She saw herself as brilliant and talented. I figured she thought the rest of us were too dumb, that we couldn't figure it out. Piper thought she was invincible."

Day four of the trial, Thursday, began with a hubbub outside the courtroom. Loni Gosnell and Jean, Piper's sister, brought in makeup for Piper to wear in the courtroom, in a bag they'd gotten in the mail from Tina. Jean handed it to Murray Janus, who turned it over to the sheriff's guard. When the guard searched inside, under the makeup and a small box containing earplugs, he discovered three Adderall pills tucked between two pieces of cardboard.

Gosnell and Jean were immediately taken to an interrogation area and questioned.

Having never had anything like this happen to him before, throughout his four-decade career, Janus was speechless. "I felt like the bag man," he says. "I told the sheriff, 'Hey, I didn't know.'"

In the interrogation room, Loni and Jean were searched, and outside in the parking lot, Loni's car was examined as well. Before being released, they were warned that passing a narcotic to an inmate could result in up to ten years in jail.

When testimony began, it would be a busy day for the prosecutors, as they put on one witness after another: from Tarra Watford and Ray Seward, from Eagle car rentals, to Tina Landrum, who testified that Piper smelled of alcohol the morning after the murder at the Miller Mart in Williamsburg. Among them were the nuts and bolts witnesses, those who brought records to be entered into evidence, explaining the Wells Fargo debit card, the receipt for the cable lock at Academy Sports, and the two wigs from wigsalon.com. In between, the prosecutors showed the videotape on which a woman who looked very much like Piper Rountree wearing a blond wig used an ATM in Virginia, the day before and the day of the murder. In the 7-Eleven tape, even the rented maroon van was visible.

When Mac McClennahan took the witness stand, Owen did the questioning. He explained his relationship with Tina, that they'd lived together off and on over a period of years. "I considered Piper a younger sister," he said.

Mac had traveled with Piper early in the week of the murder to Galveston to learn to be a landman. On the way home that Tuesday, he recalled how he mentioned wanting to go to a gun range to Piper.

"I'd like to go, too," he quoted her as saying.

At the range, Mac told the jury, Piper had rented a .22 to shoot, but ended up practicing with a .38 revolver. When Ashman asked what Piper was shooting at, an image flashed before those listening: the silhouette of a man's torso attached to a line with clothespins. Along with the gun, Mac put the Time Warner cable bill used to rent the car at Eagle in Piper's hands.

Piper had told him she'd be in Fort Worth on Thursday and Friday, and he'd worked at the clinic with Tina on Friday morning, before going to the Hill Country to see his parents. Ashman emphasized this bit of information. Later, the prosecutors would also put on one of Tina's employees, to say Tina was in Houston, not Richmond, on the morning of the murder.

When Mac next saw Piper, Fred was dead, and Piper hugged him and said, "I love you. Please don't tell the police about the gun range. It will just complicate things."

When Cari gave Piper her condolences, Mac heard her reply, "I'm sorry for the way Fred died, but I'm not sorry that he's dead."

Mac's testimony was damaging, and Janus attacked it with force. Had Mac smoked pot with Carol Freed? the defense attorney asked.

"I don't remember smoking marijuana with her," Mac responded.

"At one point you worried you could be charged?" Janus asked.

"Yes, sir," the disc jockey with his long hair anchored in a ponytail replied.

But in the end it would seem as if Janus relied on bad information, perhaps ideas he'd been given by his client.

"You're a gun freak, aren't you?" he said to Mac.

"No, sir," Mac said.

"How many guns do you own?"

"One," Mac replied.

Perhaps one of the more touching moments of the trial came when Mac said he still cared for Piper and that he hadn't wanted to testify. It would seem that both the Rountree sisters, Tina and Piper, had a strong pull on the men in their lives.

"You still had feelings for Tina, didn't you, sir?" Janus said, suggesting Mac had a reason to protect her.

"Yes, and I still do," Mac responded sadly.

That fourth day of the trial, someone was missing from the courtroom, another of Piper's siblings, Bill Rountree. He'd been a constant presence, seated next to their mother, Betty, taking notes. But the day before, police officers contended that they'd seen him sharing information with defense witnesses congregated in the hallway. Since they would testify, like all witnesses, the defense witnesses had been banned from the courtroom and were barred from hearing the evidence being presented.

Called before the judge, Bill Rountree maintained he wasn't talking about the trial but about problems at home in Harlingen. Still, the next day, Bill Rountree had left for Texas, and Piper had one fewer supporter in the gallery.

The gun evidence opened the trial that Friday, as another witness—a clerk at the 59 Gun Range—put a .38 caliber revolver, consistent with the type of bullets recovered from Fred's body, in Piper's hands. Then the jury learned Piper hadn't practiced her shooting not just once, but twice. Yet Janus pointed out that at the Sportsman's Outlet no one had been able to identify the woman who'd used the practice range, and the receipt wasn't signed Piper Rountree.

"The name of the shooter is Tina Rountree?" Janus asked.

Yes, the witness agreed.

Yet, there was more. The damning Hobby Airport records showed that Piper's Jeep Liberty was in the airport parking lot during the time she said she was in Houston. Why would her car have been logged in at the airport if, as she claimed, she'd been driving it around Houston?

Meanwhile, behind the scenes, the prosecutor's decisions were being finalized.

Many of the witnesses had given their testimony and left for home, leaving fewer in the witness room. Soon, Carol Freed was dismissed. Kizer, Ashman, and Reid had mulled over her value to the case versus her erratic behavior and decided she couldn't be counted on to remain calm while on the stand. And then there was Charles Tooke, who'd grown to realize without being told that his participation was in limbo. The other remaining prosecution witnesses had been shepherded up to a room on the second floor, but he'd been left alone in a first-floor library. With time to consider the past months and his ongoing scuffle with the prosecutors, Tooke realized why: He was no longer a witness for the prosecution but a liability to them.

* * *

As their long list of witnesses—more than fifty—drew to an end, the prosecutors brought forward two people Piper had once hoped would be her alibi: Cheryl Crider and Kevin O'Keefe.

On the stand, Crider insisted she'd never been convinced that Piper had been in the bar the night before the murder, but that Piper had brought in a notary, asking her for a statement that said she was. When Kevin O'Keefe took the stand, he looked over at the defense table and thought about how different Piper looked, heavier and pale.

Under cross examination by Janus, O'Keefe admitted that at first he believed he had seen Piper at the Volcano on that Friday night, and that it was only after checking "my invoices, etcetera, and made sure of what day it was," that he realized it had actually been Saturday, the evening after the murder.

On redirect, Kizer asked O'Keefe, "Were you at the Volcano at all on that Friday, October twenty-ninth?"

"No," O'Keefe said.

"Are you sure about that?"

"Positive."

With that, at 1:00 P.M. on the fourth day of the trial, Wade Kizer announced: "The Commonwealth rests."

As he left the courtroom for lunch, despite all the evidence he'd heard, Bill Bootwright still didn't believe Piper Rountree was guilty of murder. The father of Paxton's good friend judged the prosecutors' case as purely circumstantial. For all their records and witnesses, in his opinion Kizer, Ashman, and Reid hadn't delivered "a smoking gun" proving Piper had murdered Fred.

When court resumed that afternoon, at 1:45, Janus argued before the judge precisely what Bootwright had been thinking, that the evidence was too circumstantial and the prosecutors hadn't proven their case. He asked Judge Harris for a directed not-guilty verdict. "There's no evidence this defendant, Piper Rountree, shot and killed Fred Jablin," Janus contended.

After a moment's thought, the judge agreed, "Obviously, it's a circumstantial case." But then he ruled that the evidence, while circumstantial, was strong enough for the trial to continue.

Then, before the jury was brought in, Janus put something else into the court record, that although he and Stone had advised Piper that she wasn't required to testify, that they had made it clear that it was her absolute right not to, she had made her own decision. After a few prosecution witnesses, they said, Piper intended to testify on her own behalf.

At the prosecution table, Ashman, Kizer, and Reid all felt a sense of excitement. They had hoped for but not expected the opportunity to question Rountree on the witness stand.

With that, Murray Janus called the first defense witness, a man the prosecutors had brought to Virginia as their witness: Charles Tooke.

As Taylor Stone asked the questions, Tooke explained that he'd worked with Piper and what they, as landmen, did. He then talked about their relationship, and that he'd never known her to drink beer, which contradicted the testimony of Tina Landrum at the Miller Mart, who'd said Piper had wanted to buy a six-pack on the morning of the murder. Then Stone asked Tooke to recount the e-mail he'd received from Piper at 3:21 that afternoon. It was an attempt to show she could not have been on the airplane, as the prosecutors maintained, until 4:40 P.M. Tooke confirmed that he had in fact received an e-mail from her that day, while the plane was still in the air.

All of Tooke's testimony was important to the defense, but then came the coup de grace: Stone asked if Coby Kelley's notes on his meeting with Tooke were accurate.

Tooke answered, "Uh, I found three, I believe three things in them that I found to be inaccurate."

There it was: the assertion that the police hadn't been truthful.

On cross examination, Owen Ashman asked Tooke what Piper had responded to him after he'd replied to her e-mail that Saturday. "Well, there wasn't one," Tooke said.

"There's no response?" Ashman repeated.

"No," he said.

It was the same regarding the phone message that afternoon from Piper, in which she again said she had a question for him. "No question was ever asked of you by Piper Rountree, is that correct?" Ashman asked.

"Correct, in this regard," Tooke answered.

With that, Ashman led Tooke through a series of questions that nailed down for the jury what many of them probably already knew: that an e-mail could be sent from any computer, anywhere, by anyone. As to the squabble with Kelley, Ashman made it seem more a simple misunderstanding.

In the end, much of what Tooke testified to would be more damaging to the defense than helpful. Yes, Piper had told him that she had things on her computer she didn't want the police to see. The jury had already been told that by the time the search warrant was executed, the computer cords hung down loose, and Piper's computer tower was missing.

Despite the tumult over the Adderall, Jean took the stand as the defense's second witness. As Janus asked the questions, she explained that all three Rountree sisters were close and that she'd planned to have Callie in her wedding party the previous November. But that was unimportant. What Janus wanted the jury to hear was what Jean said next, that she'd sometimes called Piper on her 7878 cell phone and Tina had answered, and that on the telephone, the Rountree sisters all sounded much alike. "We used to fool our boyfriends, and call them up and pretend to be the other sister," Jean said.

Could the witnesses be wrong? Could it have been Tina, not Piper, on the phone in Virginia that weekend?

While Jean may have scored points for Piper, the next witness up proved to be more of a liability: Loni Elwell Gosnell, Piper's old friend. She, too, testified that Piper and Tina sounded similar on the telephone.

Yet on cross examination, Duncan Reid led her through a series of questions that ended up hurting Piper rather than helping her.

Yes, Loni admitted, she had called Piper the afternoon of the murder on her 7878 cell phone, with no response.

"You did in fact call, contact her, and speak with her at 6:21 eastern time on October the thirtieth?" Reid asked.

"Yes," Loni said.

"And the person you talked with was whom?"

"Piper."

"Are you sure you weren't talking to Tina?"

"No sir, I was not talking to Tina."

"Not a scintilla of doubt in your mind?"

"No sir."

On the stand, the white-bearded Marty McVey looked like the Santa Claus he played at a Houston children's hospital over Christmas, and he was about to deliver to Piper Rountree what many would see as a gift. To begin, Janus had McVey list his credentials: a former assistant district attorney who had once worked with the DEA. Janus asked McVey for his legal opinion about the letter to Jerry Walters. How long did a couple have to live together to be common-law married in Texas? McVey answered that there were no time limits, and that the couple simply had to present themselves as husband and wife.

Yes, he said, Piper had once given him a .38 caliber. "It

has remained in the top drawer of my dresser," he said.

Then McVey said something that could have thrown into question the linchpin of the commonwealth's case. According to the witnesses who'd testified for the prosecution, Piper was on a flight to Houston that Saturday after the murder at 4:30 P.M., but McVey maintained that wasn't possible. Because at that precise time, he said, Piper Rountree was in his office seated across from him.

"She just dropped in?" Janus asked.

"Yes," McVey answered.

"Can you tell us how you are certain today . . . that this was the time you saw her?"

"That was the day before Halloween . . . and my son was going to a very large Halloween party on that Saturday night. The party started at seven . . . He came home just a little bit before five o'clock that day."

"Could you tell us whether or not Piper Rountree was there at that time?"

"Yes," McVey said.

"Had she been there for some time?"

"Yes, sir."

On cross exam, Ashman asked, "You did not tell the attorneys, or anyone for that matter, that you saw Piper on Saturday afternoon, October thirtieth, until February, is that correct?"

"I don't know," McVey answered. "I don't recall the first time anyone asked me that question . . . if you're asking me if I volunteered the information, no, ma'am. I don't volunteer information. I answer questions."

"So when the investigators were in your office on Sunday, October thirty-first, you knew they were talking to Piper . . . about the death of her ex-husband . . . and you did not tell them you saw her Saturday afternoon? Is that true?"

"That's correct."

"Did you tell any of the police that day?"

"No, I did not," McVey said.

There was a major discrepancy in McVey's testimony and what Piper had told the court and prosecutors, however. "Well, do you know why Piper Rountree would have filed an alibi in December stating that she was at home [in Kingwood] all day on Saturday [the day McVey contended he'd seen her in his office]?" Ashman asked.

Janus interrupted. "No, he would not."

"I have no knowledge of that," McVey said.

"You are Piper Ann Rountree, is that correct?" Janus asked when his client took the stand after a short recess.

"I am."

"Ms. Rountree, did you shoot and kill Fred Jablin on Saturday morning, October thirtieth, 2004?"

"I did not."

At times, on the stand, as she had throughout the trial, Piper cried softly. She talked of her children and her marriage, saying she and Fred hadn't "seen eye-to-eye," but that by 2004 all was amicable, and that the calm had been "an answer to my prayers." She even claimed that Fred had agreed to let all three children spend their entire vacation with her that coming summer, in 2005.

When it came to her sister, Piper said she was four inches shorter and much thinner than Tina, and that Tina, not she, was the one with blond hair. Unlike Tina, she said, she had never owned a .38 caliber revolver. Tina, Piper testified, was also the Rountree sister who had a collection of wigs.

"Any blond wigs to your knowledge?" Janus asked.

"Yes." Then she went on to say that the wigs she'd ordered had never been for her, but for Tina.

Tina, too, often used her 7878 telephone, despite Piper's friends' testimony that they'd always called Piper on that number. She contended that she knew nothing about the Southwest Airlines ticket in Tina's name. The gun she'd shot at the 59 Gun Range, despite testimony to the contrary, she said, had been supplied by the shooting range. And, Piper said, she wasn't the one using Tina's identification the following day at the Sportsman's gun range. In fact, she said she'd never had Tina's ID.

From the stand, Piper insisted she'd never gone to Hobby that week, and that she'd never checked a gun with Kathy Molley. She offered no explanation as to why Allan Benestante believed he'd seen her with a .38 in her luggage. She claimed that someone else must have used the Jerry Walters debit card and that the last time she saw it was the Monday before the murder.

"Would you tell us whether or not you recall seeing the debit card after that?" Janus asked.

"No," she said. "Jerry and I had a big argument about that, too."

In Piper's version, Cheryl Crider not only recognized her from that Friday night at the Volcano but knew her by name, saying, "You're Piper, aren't you?"

Not only did Piper claim her voice sounded like Tina's, but she said they looked so much alike that their own mother couldn't tell her apart from Tina in photos. "At least black-and-white photos," Piper said.

Like Janus, McVey would later say he'd believed Piper should not have taken the stand. And in the audience, Bill Bootwright winced as he listened to her testimony. Throughout the trial, he'd been convinced there was reasonable doubt that Piper was guilty. Now, listening to her, he no longer felt that way. "Her eyes were all over the place, and she couldn't answer half the questions. She couldn't explain things, like why her car was at the parking garage," he says. "She looked guilty."

When Wade Kizer rose to cross examine Piper Rountree, he began gently, having her reiterate that the relationship between her and her ex-husband was cordial.

"When you found out he'd been murdered . . . did you immediately come to Richmond?" Kizer asked.

"I—I asked, uh . . ."

"My question is, did you come to Richmond immediately?"

"I tried," Piper said.

"Did anyone prevent you from coming?"

"Not physically, no."

"Did you go to Fred's funeral?"

"Um, I couldn't."

"Isn't it true that as of the time of his death, you weren't on friendly terms, and that it was a contentious divorce?"

"I don't think so," she said.

"You lost custody of your three children . . . wasn't that devastating to you?"

"Yes," she said.

"Wouldn't you do anything for your children?"

"I wouldn't kill for them, no."

Then Kizer zeroed in on Piper's behavior at the Volcano, her words that afternoon, that her boyfriend, not her husband, had been murdered. Throughout the trial he'd been firm yet calm, but now the commonwealth attorney's voice rose in anger.

"Why didn't you tell them the truth?" he asked.

"I didn't know where, you know, where he'd sit down and start talking to people about—"

"Is it hard to say it was my husband that got murdered?"

"Yeah."

"Because you did it. Right?"

"No," she said quietly.

Murray Janus jumped up and objected. "There's no need to shout at the witness."

Kizer pointed out that on the stand Piper had exaggerated the differences in her height and Tina's. While she'd made it sound like four or five inches, according to their drivers' licenses they were only two inches apart.

"I was trying to be accurate, sir," she said.

Piper claimed that she didn't understand the cell phone

bill records, and then said that not only Tina, but Jerry and Mac, had sometimes used her cell telephone.

"Why did you throw pieces of the telephone away?" Kizer asked.

"It wasn't working," she said.

Why, Kizer asked, if she'd been at McVey's that Saturday, had she filed an alibi with the court, an official record, not revealing that information but saying that she'd been at her home in Kingwood the entire day?

"I think you're asking two different questions," Piper answered, sounding like a lawyer parsing a question. "One is a legal alibi, and what I was doing at the time. No one asked me that. The police did not ask me that before."

If she had been at the house in Kingwood, why hadn't she answered the door that Saturday afternoon when Investigator Breck McDaniel knocked and rang the bell, even calling her house phone to let her know he was outside? Kizer asked. Piper claimed she was in the shower, and that her car was in the garage.

"Did you hear him testify there was no vehicle there?" Kizer asked.

Janus objected, and the judge sustained the objection.

Point by point Kizer skillfully went through the evidence, and Piper had little response other than to say she didn't know. When it came to the children's testimony that they'd talked with her on October 29, she said she'd stipulated to it but didn't agree with it.

"You want the jury to think that Tina committed the murder, don't you?" Kizer charged.

"I have no idea what happened," she snapped back.

"You have none?"

"I don't."

After Piper left the stand, the jury was excused, but the lawyers and judge stayed in the courtroom. Wade Kizer announced that he planned to call Paige Akin, the *Richmond Times-Dispatch* reporter who'd been covering the case, to the stand. Weeks earlier, in January, Akin had traveled to Houston to write an in-depth feature article on the Rountree case. While in Houston, she'd interviewed McVey. In the subsequent article that ran in the newspaper, Akin had quoted McVey as saying that when Piper showed up at his office followed by police officers on that Sunday, October 31, the day after the murder, it was the first time he'd seen Piper that entire year.

An attorney for the *Times-Dispatch* argued against putting Akin on the stand. Judge Harris, however, ordered that she testify, limiting the questions strictly to the matter at hand. Akin, tall with straight dark blond hair, reluctantly took the stand, and, with the jury back in the room, Wade Kizer asked her about her interview with Marty McVey.

"Did you ask him when prior to Sunday, October thirty-first, he had last seen Piper Rountree?"

"Yes," Akin responded. "He said it had been quite a while, uh, about a year, I believe."

After Akin stepped down, Janus called McVey to testify on rebuttal. Yes, he admitted, he'd only told Breck McDaniel a week before the trial began that Piper had been in his office that Saturday. But, he still insisted, she was there.

"And you deny that you told [Paige Akin] . . . that the first time you had seen her, Piper Rountree, that weekend was on Sunday, October thirty-first?" Kizer asked.

"I deny I told her that. Yes."

With that, the testimony ended.

Was McVey lying? Was the circumstantial evidence in the case enough to cancel out reasonable doubt? Was Piper Rountree guilty of murdering Fred Jablin? Closing arguments would begin the following morning, and then the jury would have to decide.

19

I t was a Saturday morning, yet Judge Harris's courtroom buzzed with activity. The jurors and attorneys had agreed they'd forgo the day off to hear closing arguments and work toward a verdict. The judge thanked them all for coming in on what was a cloudy day in Richmond, and Wade Kizer began by thanking the jurors for their attention throughout the four days of testimony.

"It's an important case," Kizer said, looking directly at the jury. "It's certainly an important case to the defendant, Piper Rountree. It's also an important case to Fred Jablin. The defendant has had the opportunity since Tuesday, with each and every day this week, to sit here in the courtroom. Fred Jablin, obviously, can't be here.

"It's easy without him here to not take notice so much of his death and not think about it. But don't ever forget that less than four months ago, also on a Saturday, at five-thirty in the morning, this man was alive . . . Two days later, on Monday morning, Fred Jablin was in the Medical Examiner's Office on a steel table and cut apart."

At that, Kizer explained the court's instructions to the jurors, the law they were to use to come to a verdict. Then he went through the long list of evidence against Piper Rountree, repeating much of what the jurors had heard in testimony, and now taking the additional step of structuring the clues to reach a specific conclusion: that Piper Rountree had murdered Fred Jablin.

First Kizer talked of the defense witnesses. None of them had really helped Piper, he argued. Charles Tooke's testimony didn't put Piper in Houston that day. As to her sister, Jean? She'd said their voices were similar, but admitted on cross exam that she could tell her own children's voices apart. "This is important," Kizer said, because two of those who put Piper on her cell phone in Virginia were Paxton and Callie, Piper's children.

Marty McVey, Kizer argued, was a personal friend of Piper's, and he hadn't told the police about his Saturday meeting with Piper until the week before the trial. Paige Akin, Kizer insisted, was an impartial observer, a journalist who didn't want to testify. "Was Martin McVey biased? Was Paige Akin biased? It's your decision," Kizer said. ". . . The evidence, ladies and gentlemen, is overwhelming. There can be no doubt that the person who murdered Fred Jablin, who shot him in the back and shot him in the arm Saturday morning, was the person who flew up here from Texas and flew back Saturday morning."

The murder, he said, wasn't random, but personal. "This killing was an ambush," he said. "It was an execution."

What did they know about Piper? "She was the mother of these three children, and she lost custody. Wouldn't that be devastating for any mother to lose custody of their children? And she lost it, she didn't get it, and Fred Jablin did."

Fred was the one who blocked Piper's attempts to declare bankruptcy. "A reason to kill him? Yes. She's desperate for money. She can't make a living as a lawyer, can't in Texas or in Virginia. She's gone to doing landman work."

If Piper were innocent, when she heard Fred had been murdered, wouldn't she have immediately flown to Virginia to be with her children? Kizer asked. She'd asked Mac not to

326 / Kathryn Casey

tell the police about the gun range and asked Jerry not to report the debit card stolen. Were those things an innocent person would do?

"Obviously, she believed that this murder was a vehicle to regaining custody of the three children, because what did she do in the days afterward? She came up here on November eighth and tried to move the court to give her custody of the kids. And instead, she got locked up for murder."

Janus, Kizer said, would tell them Piper was too smart a person to have left so many clues. "Well, she's a clever person," Kizer countered. "And she thought she was going to get away with it. She made a lot of efforts to get away with it," especially disguising herself as Tina to fly into Norfolk, rent the van, and stay at the hotel.

Piper, the former prosecutor, had also gone to great lengths to set up her alibi, telling Mac she was going to a conference, calling her children and telling them she was at work in Galveston, calling Doug McCann and acting like she was in Texas and wanted to go out to dinner, and then going to the Volcano the evening after the murder, so she could return the following week and find someone who remembered her who might be confused and willing to say she'd been there on Friday night.

"Who was on that telephone on Thursday, Friday, and Saturday? The people who testified under oath told you that Piper was," he said. "What were the reasons Tina didn't commit this murder? Well, there's absolutely no evidence that it was her."

Then Kizer brought up something he hadn't mentioned during the trial, something he wanted the jury to consider: the *T* on Tina's driver's license was written right to left. All the *T*'s in evidence, on the hotel registry, at the firing range, at Eagle renting the van, were all written left to right. And, if that didn't convince them, he reminded the jurors that two witnesses had sworn that they'd seen Tina in Houston that Friday and Saturday.

"You either must believe Marty McVey and Piper Rountree, because, again, they are the only two people who have given evidence that Piper was in Houston over the Thursday, Friday, and Saturday, or you believe the enormous amounts of evidence that shows that she was the one who flew up here with the .38 caliber revolver."

"The police absolutely have done their job in this case. I believe that the prosecution has done its job," Kizer concluded. "It's time for you as a jury to do your job. And the evidence of her guilt is overwhelming. I ask you to find her guilty of both charges."

After a ten minute recess, Murray Janus took over the courtroom for his closing. He walked up slowly and appeared tired. "What you've seen is a whole array of exhibits, a whole bunch of evidence, but what you need to think about is, is there evidence beyond a reasonable doubt that Piper Rountree shot and killed Fred Jablin? Do you have your suspicions? Sure. Do you have probables? Sure. But that's not enough."

As Kizer knew he would, Janus then listed all the prosecutors didn't have. They had no scientific evidence tying Piper to the crime. "You've heard a lot about cell phones, a lot about credit cards . . . how about fingerprints?"

The police hadn't found any, not at the crime scene nor in the rented van, not on the hotel key found in the van nor on the airline baggage receipts. "You haven't heard any handwriting expert testify that that was Piper Rountree's handwriting, not Tina Rountree's."

And the DNA they'd found at the crime scene and in the van? None of it matched Piper. Kathy Molley and the people at Eagle had all seen Tina's driver's license and didn't question that the woman who stood before them was the same woman in the picture. Others couldn't identify the woman they saw as Piper, from the clerk at the CVS pharmacy to the Papa John's deliveryman.

"Human beings make mistakes, and we're all human beings," Janus said. "The witnesses are human beings."

Always he returned to the center of his case: "You've heard, as I told you in the opening statement, the name Tina Rountree, Tina Rountree, Tina Rountree . . . The evidence is replete with Tina Rountree's name."

If Piper had been in disguise, Janus asked, why wouldn't she have traveled as a stranger instead of a person they could trace directly back to her? The identifications, he said, were faulty, because police didn't show a spread of photos of women Piper's age, only one photo, hers. Witnesses had pegged the weight of the woman they'd seen as anywhere from 125 to 140 pounds, when at the time Piper weighed only 104. Her height was five-four, but Benestante had estimated the woman he'd seen was as tall as five-nine.

Did the jury believe that Marty McVey would fly into Richmond and get on the stand and lie? McVey was an attorney, and he knew he could be charged with perjury and, if convicted, lose his law license.

"Piper Rountree didn't get on the stand and say, 'I think Tina did it,'" said Janus. "That name comes up in these documents that are in evidence."

Janus went through the court's charge to the jurors, arguing that Kizer and the other prosecutors hadn't proved their case. "There is no burden on the defendant to produce any evidence," Janus said, asking the jurors if they'd heard any of the witnesses say, "I saw Piper Rountree, that defendant, in Fred Jablin's neighborhood, in his yard, on Saturday the thirtieth of October, 2004."

"If one single one of you doesn't think the commonwealth has sustained its burden beyond a reasonable doubt, I urge you, on behalf of the defendant, Piper Rountree, to have convictions and not just go along . . . I ask you to retire and come back, following the law, to find this defendant not guilty."

* * *

As the state has the heavier burden throughout the trial, they are allowed to split their time in the closing statement, to sandwich the defense. Kizer had chosen to do that, and now Duncan Reid took over. Throughout the prosecutor's office, Reid was known to be an exceptional narrative speaker, one who could crystallize for a jury the essential elements of a case.

First Reid restated what they knew from the cell phone bills: Yes, he said, they did know that Piper had been the one at the Sportsman's gun range. How? Because just minutes before, she'd called 411 on her cell phone to get the gun range's phone number. And at the Sportsman's, there was no conflicting testimony, the woman who'd practiced brought her own .38.

Yes, Reid admitted, some of the identifications were off by height and weight, but, he said, the witnesses were human beings and not computers. They recognized Piper as the woman they saw, and that was all that mattered. How in the world, he asked, could so many who said they saw her be wrong? How in the world could so many who said they'd heard her voice on the cell phone be wrong? Could Mr. Janus explain that?

On the stand, Piper had testified that she'd talked with her children that Friday, but she said that she'd used Tina's telephone. Tina's cell phone records were in evidence, and Reid suggested the jurors look through them to see if they could find Piper's calls to the children in Richmond. "You're not going to find a single phone call from Tina's cell phone into the 804 area code," he said.

On the stand, he said, Piper Rountree had told one lie after another, "spun and spun and spun like a car out of control."

The crime that ended in Fred Jablin's murder, he said, began when Piper deceived herself by thinking, "I can do it, I can do it and not get caught," Reid said. "She deceived herself by thinking . . . the kids are going to be better off with me than they are with him. They won't miss their daddy."

Then, Reid used the bit of evidence he'd elicited during the trial, that during the camping trip Piper, a cell-phone-aholic, hadn't made a single phone call. Why? "She had a grand time," he said. "She was enjoying those kids."

On Monday, when she had to return them to Fred, Reid said, her heart sank, and she knew she had to get them back. How? Get rid of Fred.

Reid then recounted in chronological order the events that led to the murder: from buying the wigs on October 21, to trying to purchase the airline ticket on October 25, to practicing at the gun ranges on October 26 and 27.

During an emotional moment in his closing, Reid showed the jurors a recent photo of Fred with the three children, all smiling and happy. "This is the sort of family we all hope to have, the sort of family he had, the sort of family she wanted."

On the Friday evening before the murder, Piper went to sleep, her children went to sleep, and Fred went to sleep. "At four-thirty Saturday morning she checked her voice mail, summoned her courage, got her gun, went out into the night, drove down the road she knew so well, went through the neighborhood she used to live in. Her children slept. And Fred slept.

"She parked her car, got out, and stepped out into the night. Took a position. Her children slept," Reid said, in a hushed, dramatic tone. "Fred got up, put on his slippers, went down-stairs, started the coffee, went outside, walked down the driveway"—here, Reid's voice sped up, tight and tinged with the horror of that morning. Fred "saw her, turned, ran, got shot in the back, one, two, three bullets, lay, fell, dead, under the basketball court that his children played at. She ran away into the night, and her children slept.

"The police came and found Fred early in the morning.

Then the police went into the house, awoke the children to the nightmare that their own mother had created. Jocelyn, Paxton, and Callyn lost their father that day, all because she wanted to be with them.

"A murder in the first degree, willful, deliberated, premeditated murder of Fredric Mark Jablin."

The two alternates were then dismissed, the twelve remaining jurors left the courtroom in single file to deliberate, and the lawyers left to find a place to have a quiet lunch. The courtroom emptied, and many of those who'd come simply as spectators went home to spend the rest of the day with their families, perhaps feeling lucky that they were still together, whole.

No one could have predicted how long the jury would be out. It could have been minutes, hours, days, or even weeks. But the jurors went for lunch, and then, less than an hour after their deliberations began, at 2:45 that afternoon, word circulated through the Henrico County Courthouse that the Rountree jury had reached a verdict.

It had all happened so quickly that Piper's mother and friends had not returned to the courtroom in time to hear the verdict. As the judge read, Piper Rountree rested her head on her hands and sobbed. The jury had found her guilty on both charges: first degree murder and the use of a firearm during the commission of a crime.

The judge asked the clerk to poll the jury, and she did, calling out each juror's name. "Is this your verdict?" she asked.

"Yes," each answered.

Through it all, Piper cried.

The jury's work hadn't ended. Now they were charged with one more task: to make a sentencing recommendation to the judge.

They waited briefly for all the witnesses to arrive, and the first to testify in the punishment phase was Betty Rountree, Piper's mother. She talked of Piper the baby of the family, whose father had suffered a massive stroke at forty-eight, one that devastated Piper. "He was really lost in this world," Betty said. Yet, she stressed that Piper had never given her mother problems. She'd been a good daughter, a bright child, and an excellent student who'd loved to read. And she was an artist who painted watercolors.

"Her work is just beautiful," she said.

On the stand, Betty Rountree looked tired and defeated as she told the jurors how devoted Piper was to her children. "Her whole life was the children," she said. "Taking them places, reading to them. And the children were good children."

Janus used Betty to bring into evidence photos of Piper with the children, during happy times. One was of Paxton hugging his mother, another was with Callie, when she was a baby, and Piper was kissing her toes. In another photo, Jocelyn had her head on Piper's shoulder.

"Is there any question in your mind that Piper Rountree loved those children?" Janus asked.

"Oh no," Betty said. "I don't think there's any question in anyone's mind that knew Piper or knew the children. She was . . . she was just one of the best mothers that you could imagine."

Then Betty recounted a night she'd stayed with Piper and the children. They'd all gone to bed in their individual beds, but in the morning she'd found all three children had slipped into Piper's bed during the night. That was how Betty Rountree found her daughter and three grandchildren, all asleep together, in each other's arms. "It was the most peaceful thing I've ever seen."

When Wade Kizer took over, he used Betty to illustrate how supportive Fred had been of Piper, not only sending her to law school after they married, but moving to San

Antonio and commuting three hours so they could live there.

"Ma'am, it was devastating for Piper to have her father taken away [after the stroke]?" Kizer asked.

"Absolutely," she responded.

"But he was taken away by a medical condition . . . and she had the opportunity to see him until she was approximately thirty-five years old?"

"Yes," Piper's mother admitted.

With that Loni Gosnell took the stand in what would be the strangest testimony of the trial. Throughout, Loni sobbed, describing Piper as a wonderful mother, a Brownie leader and an art teacher.

"She's wonderful," Loni said. "She's my best friend."

Then, suddenly, for seemingly no reason, she turned to the jury, angry. "Look at me when I'm talking to you," Loni ordered. ". . . They're not looking at me!"

The judge admonished her, asking her not to talk directly to the jury.

When Duncan Reid took over, he asked Gosnell if it were true that she'd visited Piper six or seven times while she'd been in jail. Gosnell said yes, she had.

"Did you attend Fred Jablin's funeral?" he asked.

"No, I did not. I didn't know about it, honestly."

"Have you visited his grave?"

"No, I have not."

The prosecutors put on the stand only one witness, Michael Jablin. He talked of the Fred he'd known, his brother and his friend. And he described the horror of that weekend, when he and his wife had driven to Richmond to pick up Jocelyn, Paxton, and Callie.

"They're all seeing therapists now," he said, giving an inkling of the damage the divorce and murder had done.

When he'd come for them, they'd had to move quickly,

and the children were forced to leave behind not only their home, their friends and neighbors, but even their clothes and toys. "It's been very traumatic for them," he said, wiping away a tear. "It's not an easy experience when you lose your father, and you have to move very quickly, and you can't go back to your own home . . .

"Well, Jocelyn, every night almost, she sits in her room and she cries because she misses her father so badly. It's been very hard to explain what happened. You know, when her father was murdered, you know when you know who it was, I said to her, we have to find the facts out first . . . she worried about my wife and myself, if someone would murder us . . . I dread going back now with my wife and having to tell them this is what the outcome is . . . How do I explain to young children that their mother killed their father? They've lost both parents now . . . How does anybody explain something like that?"

Michael Jablin said that he'd asked Paxton and Callie to write their mother, but neither wanted to. "These children are very close . . . they don't want to talk about what's happened. We have therapists, psychiatrists working with them, to try to help them understand," he said. "This is a difficult process. It's going to take years, and into their whole life to understand what happened here."

After more thought, Michael added: "It's hard for me to understand why a mother would do this to her children."

There were more photos of Fred and the children, happy days when they played basketball and at a party Fred had thrown for Paxton, depicted in a photo of Paxton blowing out the candles on his birthday cake.

"Mr. Jablin, don't you think the children are going to miss their mother?" Janus asked on cross.

"There's no question. They're going to miss both parents."

"Jocelyn came to visit her mother, after Christmas, right?"

"Correct," he said, later adding that on the advice of a grief counselor he had photos of both parents with the children displayed in their home. As to the money, Fred's estate and the insurance money, he said that had all been put into a trust fund for the children.

Finally, in what would be the prosecutors' final argument, Owen Ashman took over the courtroom. From the start, she pushed the jury to give Piper the maximum, life in prison. Fred, she said, was a teacher and a devoted father. He'd been killed outside his own home, in his slippers. "Left there to be found by neighbors, perhaps his children, who were sleeping in the house.

"Think of the three children," she pleaded. "Think of how these three children must ache every single time they're asked normal daily questions by other children, such as, are your parents coming to this event? . . . Every single Halloween. He was killed the day before Halloween . . . a children's holiday, they'll think of their dad.

"You heard their mother was devastated about losing her father for medical reasons, yet Piper, the same child who was devastated . . . has taken her own children's father from them. This is not a good loving mother. No good loving mother would do this to her children."

Then Ashman's voice took on a hard edge. Piper Rountree is "clever. She's sneaky. She's selfish. She's dangerous. She's a murderer with no remorse," Ashman said, pointing at Piper. "Three shots, she took on October thirtieth. Three little lives forever changed. One life gone forever. She deserves life in prison.

"Do it for Fred. Do it for the children . . . Do it because it's the right thing to do . . . Sentence her to life."

In his closing, Murray Janus talked in a hushed, quiet voice. "It is a tragedy for the children because they lost their father," he argued. "It's a double tragedy now because they're going to lose their mother for a very long time.

"I don't think there's any question in anyone's mind that Piper Rountree dearly loved those children . . . There's no question they loved her in return. She was their primary parent . . . And what she's done is out of frustration, disappointment, devastation because of losing these children . . . a special relationship between a mother and her children. Don't take their mother away . . .

"Use your common sense," he pleaded. "Understand."

Since Piper's case didn't have the special circumstances that Virginia law required to make it a death penalty case—such as the murder of a law enforcement officer, multiple murders, or murders coupled with additional charges including rape, abduction, and extortion—Piper's possible punishments ranged from twenty years to life. While the jury was out deliberating their recommendation, Janus, Kizer, and the judge pulled out their calendars. No matter what the jury suggested, the final decision on Piper's punishment would be up to Judge Harris. A presentencing report would have to be compiled for yet another recommendation to the court on a proper punishment. The date they chose for that final hearing, the sentencing of Piper Rountree, was May 6.

Then, the judge, the prosecutors, the defense, and Piper all settled in and waited for the jury. It didn't take long.

" 'We the jury, having found the defendant guilty of first degree murder, Fredric Jablin, fix her punishment at life in prison,' " Judge Harris read.

A murmur went through the courtroom, the jury was thanked for their service, and then the judge dismissed them.

Later one jury member would look back at the decision and say they had all quickly agreed they would sentence Piper to life. Why? To benefit the Jablin children. "We didn't want Piper Rountree to ever have the opportunity to come back into the lives of those children," he says. "We made the decision the children were better off without her."

Would Judge Harris agree? Three months remained before that question would be answered.

20

As the sentencing hearing neared, Piper worried. What would she say to the judge, what would she argue to convince him to go against the jury's recommendation? If he didn't, she'd spend the rest of her life in a Virginia prison.

Piper knew what she wanted to say, that the jury was wrong, that she was innocent, that she didn't kill her ex-husband, leaving her children virtual orphans. She wanted the judge to know that her husband wasn't the perfect man, the kindly professor he'd been portrayed. Yet that wasn't what the judge wanted to hear. Now that she'd been convicted of murdering her ex-husband, the father of her children, the judge would be listening for signs of remorse.

The morning of Friday, May 6, 2005, more than two months after the end of the trial, was a somber one, a late spring storm drenching much of the East Coast. Inside the Henrico County Courthouse, both families collected in hallways where gray clouds blocked the sun from entering skylights, giving an eerie, quiet, somber effect. When the doors opened just before the eleven o'clock start time, the media filed in to fill the left quarter of the courtroom. Television cameras whirred in the now empty jury box, and notebooks were poised. In the center aisle, behind prosecutors Kizer, Ashman, and Reid, Michael Jablin and his wife, Elizabeth, sat together holding hands. He stared at Piper, appearing furious at all he'd heard at the trial.

In front of the Jablins sat the investigators, Jamison and Kelley, there to see the case through. Neither wanted to miss this final moment, the end of a case they'd pulled together from cell phone records and shreds of paper, receipts that tied Piper Rountree to Virginia and Fred Jablin's murder. Also in the courtroom were students from the University of Richmond, hoping a man they'd respected was given justice.

On the far right section, Piper's family and friends collected anxiously. Two were notably absent: Jean, who'd given Janus the makeup kit with the Adderall, and Bill, who'd been questioned about coaching witnesses. They'd been replaced by Bill's wife and Tina's son, Mike Gano, there to support Piper and her mother Betty.

Just after 11:00 A.M., Piper entered, accompanied by two uniformed deputies. She'd lost the excess pounds she'd gained pretrial, when witnesses had to identify her in the courtroom. She wore the same navy pantsuit with double-breasted brass buttons as throughout much of the trial, only now she was a convicted murderer, and when she entered, her hands were tethered behind her in handcuffs.

As a deputy unlocked the cuffs, she looked over at her mother and smiled, as Betty cried softly. Then Piper took her place next to Taylor Stone at the defense table.

The first order of business was the presentencing report prepared for the court. Janus rose and addressed the court, the elder-statesman-like man offering his objections. He looked tired and beaten. The case had been a difficult one, with so much evidence against his client, and she had tied his hands. Maintaining her innocence had cost them dearly, closing off any argument that could have formed extenuating circumstances and cut her sentence.

"On page one," Janus began calmly. His client hadn't gone

to the Volcano to set up an alibi, he argued again, merely to
see if anyone remembered her. He proceeded to object line by
line to everything he considered a discrepancy in the report.
Many were small points, that Piper's father had suffered a
stroke when she was only six, leaving her devastated, implying that this had affected her entire life. She couldn't take
hormone replacement, he said, because she had no spleen to
process the medication. Then there was the main issue. Janus
wanted it noted that at the motion hearing three of the witnesses hadn't initially recognized Rountree. They couldn't
identify her.

"That's in there," the judge said. Janus nodded. He knew
it was.

All the while, Piper whispered to Stone, as if urging him
to interrupt Janus and to tell him to say more. Janus didn't
listen. Perhaps he knew what was important even if she
didn't. He was setting up the grounds for the appeal. This
trial was over and his eye was on finding a way to set it aside
and secure her a new day in court, another chance to fight
the charge of murder.

After Janus sat down, Kizer rose. He had his own issue to
raise, one Janus had already conceded. Piper had implanted
a lie in the presentencing report, one easily discredited.
She'd claimed that her ex-husband had been arrested for
felonies in New York, Colorado, and Virginia. A search of
the records by both prosecutor and defense attorney had
found no such record for the deceased professor. Fred Jablin's only arrests had come at the behest of his wife, when,
in the months preceding the divorce, she claimed he'd
abused her, charges dropped after police found no evidence
and she failed to show up at a hearing.

"There is no such record," Wade Kizer said, frowning at
the judge.

Even now, with the trial over, Piper was lying, and about
something so easily discredited. It was as if she couldn't

help herself. Perhaps she no longer knew the difference between the truth and her complicated web of lies.

The report corrected, Janus called his first witness, Lavon Guererro, Piper's childhood friend. "Piper is a tremendous homemaker . . . she was one hundred percent there for her kids . . . a beautiful, gentle spirit," she testified. "Since we were children, she talked about having children, wanting a family."

After eliciting testimony from Guererro that showed the two women had lost touch after Piper moved to Virginia, Kizer barely bothered to question her, allowing Janus to move on.

Second up for the defense, Annie Williams, Jocelyn's soccer coach. She, too, drew the picture of Piper as a supermom, the one who stood cheerleading her kids from the bleachers during soccer games and gave the Brownie scouts arts-and-crafts lessons.

"She gave her career up to be with her children," Williams said.

"She didn't pass the bar in Virginia, did she?" Kizer corrected.

"I don't know," Williams admitted.

Next was Betty Rountree. On the stand, she, too, talked about her daughter in glowing terms, the delightful child who'd become the dedicated, loving mother. "Was she ever the same after she lost custody of the children?" Janus asked.

"No. Never," Betty said. "You don't take children away from a mother. They have a special bond."

It was a risky course for Janus. He was trying to humanize Piper, to show she'd been devastated by the custody battle, perhaps explaining in a small way what happened. Of course, the testimony also bolstered the prosecution's claim of motive, that Piper had killed to get back her children.

The defense's final witness was their most effective, Mike Gano, Tina's son and Piper's nephew. She'd been a loving aunt, he testified, one who'd played with him on the floor as a child, who introduced him to the beautiful things in life, art and drama, passions he'd carried with him into adulthood. Visiting Fred and Piper in Austin, he said, had been a happy time, a house filled with laughter, love, and nature. Piper made the world around her beautiful, he said, and she deeply loved her three children.

"How did the children feel about their mother?" Janus asked.

"They loved her," he said. "They were emotionally and spiritually inseparable."

Mike Gano appeared a son anyone would have been proud of, preppy and proper, and deeply devoted to his aunt, grateful for all she'd done for him.

"Will you be there to support her now?" Janus asked.

"Yes," he said. "She was always there for me."

With that, the defense portion rested and Wade Kizer took center stage. He called no witnesses to the stand; instead he reminded the judge of the events leading up to the murder. Piper had prepared for Fred's murder for more than a week, planning it carefully, taking steps to hide her identity and give the impression her sister, Tina, had been the one to fly to Richmond to gun down Fred Jablin.

"At any time she could have turned back, and we wouldn't be here now," Kizer said. "But she didn't . . . three shots, one, two, three, and Fred Jablin was cut down . . . it was an execution . . . she shot the father of her own children."

Piper was intelligent, well educated, with a supportive family. She had much going for her. Yet she chose murder. "And she's shown no remorse," he said. "The three children are as much victims as their father. They lost their father. Their mother to prison. If they live to be eighty years old their lives will be forever changed."

After her attorney made a few more brief comments, urging mercy for his client, Piper rose to address the court. In a hesitant voice, her hands shaking, she rambled, reading from sheets of paper she had on the table before her. "If someone had asked me months ago, I would have told them my children need a father," she said in a tearful voice. Her face flushed, she read on. "I've always maintained the children need a father. They also need a mother."

She talked in circles, saying over and over again that she would never have taken their father from her children, that she loved them, implying that taking her from them would only harm the children further. She gave up her career for them. Implicit was the idea that she had given up everything to get them, killing their father, but in the process losing them forever.

"I have been punished already," she said, casting herself as the victim. "The loss of my reputation, my assets, my children. I request for the children compassion and mercy."

Over and over she said it, just as she had throughout the contentious years of the divorce, "for the children . . . for the children . . . for the children." Yet no one in that courtroom could have doubted that in her quest to reclaim them, Piper Rountree had dealt her own children the most devastating of blows.

With that she sat down, folding her hands before her face as if she were praying.

Finally, it was Judge L. A. Harris's turn. Piper never looked up at him, but he sized her up squarely.

"In any first degree murder case there's an element of premeditation, willful, deliberate," he said. But he went on to say that in this case it went beyond that, taking in a full week of planning. He didn't replay the trial for those who'd gathered, knowing they'd all heard of the wig, gun, lock, and fake identity before. Instead he went on. "The victim, Fred Jablin, was not only an ex-husband, he was a father . . . if there were ever equal victims in a case it's the three children . . . at home sleeping as their father is gunned down in their own driveway."

A look of revulsion on his face, the judge continued, ". . . I am not going to do the easy thing in the case. I intend to do the right thing."

Undoubtedly knowing what was coming, Piper appeared to steel herself, her back rigid. Then the judge delivered her sentence: life in prison. Her first possibility of parole would be in twenty years. Piper cried, and her family and friends clasped hands, tears streaming down their cheeks.

It was over.

Minutes later Piper Rountree, attorney, artist, and self-described loving mother, was again handcuffed. She held her head erect, her face blank, as she was led from the courtroom through a back door, behind which she'd disappear for the rest of her life into the Virginia prison system. Despite all the evidence against her, she had never said the one thing that might truly have made a difference, what so many had waited to hear: that she was sorry.

Author's Note

. . . .

Two weeks after the sentencing, I drove to the Henrico County Jail East, an hour outside Richmond, to interview Piper Rountree. I was led through a maze of doors and thick interior windows, into a sunless, fluorescent-lit room. After the heavy clank of doors locking behind me, Piper arrived, so diminutive in her denim prison garb that she resembled an adolescent child. Her hair was perfectly combed and fell thick about her face, her nails carefully manicured, filed into perfect half-moons. We shook hands, and she sat across a narrow table from me, while a uniformed guard leaned against a wall five feet behind her, watching.

"No, I didn't kill Fred," Piper said, her voice high and strained.

When I asked her who might have, her eyes flashed and she became animated. As in the trial, she held her hands in front of her as she talked, palms clasped, as if praying. She told an elaborate theory that day. She said that Fred was an evil man, one with many enemies. "Anyone could have murdered him," she said. Then she repeated many of the claims she'd espoused to Coby Kelley in the week following the murder: "Someone Fred was dating, who he treated like he'd treated me, abusing me, could have done it. Maybe Michael Jablin had it done to get Fred's money. Or it could have been someone at work."

In Piper's twisted version, Fred Jablin wasn't simply a

professor at the University of Richmond, but an influential man with his hand on the pen that wrote the checks. He had, she said, access to all the university's money, the more than a billion-dollar trust. When I remarked that this seemed more than unlikely, that such funds are typically administered by trustees and university presidents, not professors, she called me "naïve," describing Fred as all-powerful, an expert in the power of persuasion. "He specialized in advising people in how they could control others through communication," she said, her brown eyes narrowing.

She went on to speculate that someone at the university could have framed her, by hacking into the computers of large corporations, including Southwest Airlines and Wells Fargo Bank, to forge records and make it appear that she'd flown to Richmond the weekend of the murder. Why? It could have been over money, she charged, or because Fred was having an affair with a UR higher-up, perhaps not a woman but a man. "It was information that would have shaken the university," she said, her voice rising. "That's why they had him killed, to shut him up and keep the truth from coming out."

She couldn't explain all the witnesses who'd seen her in Richmond that weekend, and when I asked for evidence, anything at all to back up her claims, Piper simply smiled then shrugged, offering nothing.

In truth, at UR many still grieved for Fred Jablin. The school had set up a scholarship fund for the children, a section of the faculty lounge in Fred's memory as a library that would house his work, and, something many believed he would have been particularly proud of, an award named after him for outstanding Ph.D. candidates. "We'd like him to be remembered at the university," said Dean Ruscio. "He was important to the university, and we will continue to miss him."

That afternoon, Piper and I talked about many things, from her childhood and family to her years with Fred and her love of her children. We talked about angels, and I asked her about the dark figure on her business card, the one that looked more like a demon. "I painted that myself," she said, smiling. "Angels have always been special to me."

Under Virginia guidelines, the first time Piper would become eligible for parole was in the year 2020, when she turned sixty. When the subject came up, she said earnestly, as if it were the most foreign of possibilities: "I don't understand how they can do that to my children. They need me."

When I asked about Tina, Piper's expression softened. "Could Tina have killed Fred?" I asked. Of course, there'd been ample testimony that Tina had been more than a thousand miles away, in Houston, the morning of the murder. Still, Piper wasn't ready to abandon her ruse. She smiled.

"Maybe," she said. "I really can't say. All I know is that I didn't."

I found myself staring at Piper, wondering how she could say she loved Tina and remain so eager to cast her as a suspect in a cold-blooded execution.

A month later I waited at the pool beside the Houstonian for Tina to arrive. As usual, she came late—two hours late. When she walked through the wrought-iron gate, many looked up to see the chesty blonde dressed in a leopard swimsuit and matching pants, wearing a straw cowboy hat and thick red lipstick. She carried two large bags she plopped down onto the round patio table between us. "I brought my mail and my paperwork from the clinic," she said. "I thought I'd sort it while we talk."

With that, she turned to walk over to a nearby trash can, to throw out an empty water bottle. As she passed me, she sighed, saying, "You know, some mornings I wake up and I still can't believe Piper did this."

It sounded like an admission, a confirmation that she knew her sister was guilty, that Piper had been the one who pulled the trigger, gunning down Fred Jablin in his driveway. But when Tina, who still had felony charges pending against her in Houston for tampering with evidence, returned to the table and I asked her, "What exactly do you mean?" she demurred.

"Oh, Piper couldn't have done it," she said with a smile. "Fred deserved it, but she wouldn't have done it. My sister would never hurt anyone."

We spent two hours together, baking under the scant shade of the umbrella, on a hot summer day. Tina was worried, nearly frantic about what lay ahead. "If I'm convicted, I'll lose everything," she said. "They'll take away my license, and I'll have to shut the clinic.

"Where will all my patients go?" she asked. "Who will care for them?"

As she'd once written a forty-two-page treatise on Fred Jablin, labeling it dangerous to allow him to be around children, on this afternoon Tina psychoanalyzed those who'd turned against her, her former lover Mac McClennahan and her once close friend Carol Freed. She said they lied. Why? Cari out of some confusion, which she said arose from a troubled childhood. "Cari's a mess," she said. "I don't know how she'll survive without me. She counted on me for everything. A very sad case."

When it came to Mac, she had another theory. Perhaps Mac killed Fred and framed Piper to get her out of the way, Tina conjectured. "Now he has Piper's old job. That's important to Mac."

When I asked if she thought Mac—a man she'd only moments earlier described as nurturing and kind—would murder for a job, Tina just smiled. "I can't tell you what anyone thinks is enough to kill for," she said. Of course, there was a major problem with her theory: Police had found absolutely no evidence Mac was involved, and had, in fact, through records and witnesses, determined that at the time of the crime he was in the Texas Hill Country visiting his parents.

The entire time we talked, Tina sorted through the large black vinyl bag and the duffel bag she'd brought with her, organizing bills, tearing some up, filing the others into piles: bills to be paid, credit card offers, bills for the clinic, and bills for the house.

"Mac was really good at taking care of all this for me," she said wistfully. "He was organized."

When I finally got up to leave, she rose to walk me into the hotel. While many nowadays fear identity theft and guard their information, Tina left the stacks of mail sitting open on the table next to the pool.

"You know, they don't even have a law to charge me under in Virginia, that's why they're doing it in Texas," she said as we neared the hotel elevator. Looking straight into my eyes, she then said something that for the second time that day could easily be interpreted as an admission that Piper murdered Fred, and this time something else, that Tina had, indeed, helped Piper cover it up.

"In Virginia, they understand that a family member is going to help in a situation like this," Tina said. "They know that you can't do anything less for someone you love."

With that, we shook hands and I left.

On the drive home I considered the ties that bound these two sisters, recalling the words of one of Piper's old school friends: "It was obvious that Tina was there for Piper, and Piper was there for Tina." Unlike Piper's love of Tina—one she was willing to sacrifice for her own ends—it appeared Tina's commitment to her younger sister had never faded, so much so that she was willing to risk her career, even her personal freedom, to stand behind Piper.

The following months were busy, as the summer burned away while I worked on this book. For the most part, things slowly fell in place. But gradually something remarkable happened, as I received e-mails on my website from men across the country who said they were writing to Piper in prison.

Most had seen her on television, and something about her—perhaps her stricken, waifish courtroom demeanor—touched them. The first, a Richmond actor, e-mailed that he felt drawn to Piper, that they wrote poetry back and forth, and that he was raising money for her appeal. Perhaps one day I could write the story of their love, he e-mailed, suggesting his relationship with Piper was evolving into more than friendship. It seemed even in jail, Piper's allure hadn't diminished.

On Friday, November 4, 2005, just days after the first anniversary of Fred Jablin's murder, I saw Tina again, this time in a Houston courtroom. She was there to plead guilty in a plea bargain her high-profile attorney, Kent Schaeffer, one of the most sought after criminal attorneys in Houston, had negotiated with the assistant district attorney on the case, Kari Allen, who headed the prosecutors in the 388th District Court. As usual, Tina arrived late and in a flourish, wearing a tight blue plaid suit over a low-cut blouse, glasses dangling low on her nose. As she waited for the judge to arrive, she played with her cell phone, attempting to take photos of the room, as if for a scrapbook. She seemed oblivious to what was happening.

Just after eleven-thirty, she finally walked to the front of the courtroom and stood before the judge at the bench. The plea bargain was a good one for her. The charge was now attempted tampering with evidence. Schaeffer had negotiated it down to a misdemeanor. When the judge asked how she would plead, however, Tina still appeared uncertain. She looked at her attorney, who nodded. Only then did she whisper, "Guilty." Her sentence: nine months of deferred adjudication, a $300 fine, and eighty hours of community service. If she remained out of trouble during that time, her record would be swept clean. She'd keep her license and the clinic.

Prosecutors, Allen said later, felt the sentence "fit the evidence." Unsaid was that they had little against Tina except the word of Carol Freed, who could be considered an accomplice and therefore potentially untrustworthy. But the bottom line was that Allen, like Owen Ashman in Virginia, doubted that Tina was involved in the actual murder. "We never had any evidence that tied Tina to the murder or the planning of the murder," said Allen. There is no statute of limitations on murder, and if she found such evidence, Allen said they could file further charges. But the way she added that information gave me the impression she didn't expect such evidence to emerge.

In the end, for all the speculation in Texas and Virginia about Tina Rountree's involvement in the murder, it appeared that Piper had acted alone.

Before I left the courtroom that afternoon, one of Tina's friends, Glenda King, a disheveled, gray-haired woman, a nurse, summed up much of what I'd heard during the year I'd investigated the case: "One thing about the Rountrees is that their children are more important to them than anything else in the world. I've never seen people who idolize their children the way Tina and the rest of the family does. They'd do just about anything for the children."

For the children. That phrase had been repeated so often by Piper that it had become the chorus in her tragic opera.

Months later, on the day before Valentine's Day, February 13, 2006, nearly a year after the end of the trial, Piper was once again in a Henrico courtroom, this time with Elizabeth and Michael Jablin sitting at the opposite table. They were there to finish what Piper had begun the day she took aim at Fred and pulled the trigger. She'd thought that by killing him she could reclaim her children. What she'd actually done was made them orphans and given them away.

At the hearing called to sever her parental rights, Piper appeared relaxed, dressed in a white front-tie shirt and navy pants, her hair long and cut in bangs. From the stand, Michael Jablin testified about the children. Jocelyn, by then sixteen and a high school junior, had two new cats and had begun planning her future, looking at colleges. Paxton, thirteen, and Callie, ten, were into soccer, and she was considering branching out into lacrosse. All the children were doing well in school, a portion of the credit due to the therapists who, Michael Jablin said, had "helped them so much."

"It's been a struggle to keep them moving on with their lives," Donna Berkeley, the children's court-appointed guardian, told the judge.

From jail, three to four times a week, Piper still wrote her children letters. Yet the children seemed to be separating from her. Since the trial, none had gone to see her or talked with her on the telephone. In the past year, Piper's children had written only twice: on Mother's Day and Christmas.

When the judge ruled, Piper was stripped of all parental rights, and Michael and Elizabeth Jablin officially became the children's parents.

In northern Virginia, the Jablins had bought a larger house for their expanded family. Many in Richmond and at the University of Richmond believed Michael and Elizabeth would give the children the understanding and love they needed. But still, could Jocelyn, Paxton, and Callie ever forget that horrible morning, the dawn their mother gunned down their father while they slept? If not forget, could they ever accept it?

How could they? For the rest of their lives, every Halloween, pumpkins, scarecrows and trick or treat would recall the horror of October 30, 2004.

In the prison that afternoon we met, Piper had told me: "No one loves a child as much as a mother. It's a special love, an overwhelming love, greater than any other love."

Yet, was Piper's obsession with her children really love?

I thought about the passage from the Bible, 1 Corinthians, the one read at nearly every wedding. Love, it says, is patient, kind, without envy and pride, never rude, and not

easily angered. It keeps no record of wrongs, always pro-
tects, trusts, hopes, and perseveres. "If I speak in the tongues
of men and angels, but have not love, I am only a resounding
gong or a clanging cymbal," it reads. If I "have not love, I
am nothing."

If she'd ever known, somewhere along the way, Piper had
forgotten what it means to truly love. She loved no one more
than herself: Not her sister, who she remained willing to
sacrifice, nor Jocelyn, Paxton, and Callie, whom she'd self-
ishly robbed of their father.

Ultimately, no matter how often Piper had spoken of angels, the voice that guided her the morning she murdered
Fred was far from heavenly. In the end, Piper Rountree was
seduced by greed, revenge, hatred, and evil.

Acknowledgments

- -

This was a particularly challenging book to research and write, requiring tracking down sources in Texas, Virginia, New York, Florida, Michigan, and Wisconsin. As with my first three books, Jim Loosen at JAL Data Services in Seattle worked wonders. He was tenacious even when the trail grew cold, and his help made all the difference. Thank you, Jim, more than I can say.

Secondly, thank you to my readers: Claire Cassidy, Pamela Guthrie-O'Brien, and Terry Bachman. Your input was insightful and much appreciated. Special thanks to Connie Choate and Sandy Sheehy for their continued encouragement. To Eleanor Richardson of Eleanor's Legal Support in Los Angeles for tracking down Florida records.

Many thanks to my editor at HarperCollins, Sarah Durand.

My gratitude goes out to Melody Foster in Richmond, Virginia, and John Daly, at the University of Texas in Austin. You were both patient and kind, and your guidance was invaluable.

Lastly and most importantly, thank you to the many readers who've discovered my books, enjoyed them, and taken the time to recommend them to others. There's been little publicity for my work, and your kind endorsements have made a world of difference, enabling me to continue what I so enjoy, zeroing in on a fascinating case and dedicating myself to understanding the who, what, how, and, most important, why.